AF183821

Clinical and Methodological Issues in Cardiovascular Psychophysiology

Edited by

A. Steptoe · H. Rüddel · H. Neus

With 39 Figures

Springer-Verlag
Berlin Heidelberg New York Tokyo

Dr. Andrew Steptoe
Department of Psychology, St. George's Hospital
Medical School, Cranmer Terrace,
GB-London SW17 ORE

Dr. Heinz Rüddel
Medizinische Universitätsklinik, Innere Medizin,
Sigmund-Freud-Str. 25,
D-5300 Bonn 1

Dr. Hermann Neus
Medizinische Universitätsklinik, Innere Medizin,
Sigmund-Freud-Str. 25,
D-5300 Bonn 1

ISBN-13: 978-3-642-70657-8 e-ISBN-13: 978-3-642-70655-4
DOI: 10.1007/978-3-642-70655-4

Library of Congress Cataloging in Publication Data. Main entry under title:
Clinical and methodological issues in cardiovascular psychophysiology.
Based on a meeting held in May 1984 in Linz, Federal Republic of Germany, and organized by the Commission of the European Communities' Concerted Action on Quantification of Parameters for the Study of Breakdown in Human Adaptation, in conjunction with the Dept. of Internal Medicine, University of Bonn. Includes index.
1. Stress (Psychology)–Testing–Congresses.
2. Cardiovascular system–Psychophysiology–Congresses.
3. Cardiovascular system–Diseases–Diagnosis–Congresses. I. Steptoe, Andrew. II. Rüddel, H. (Heinz), 1950-. III. Neus, H. (Hermann), 1953-. IV. Commission of the European Communities. Concerted Action on Quantification of Parameters for the Study of Breakdown in Human Adaptation. V. Universität Bonn. Department of Internal Medicine. [DNLM: 1. Cardiovascular Diseases–psychology–congresses. 2. Psychophysiology–congresses. WG 100 C6403 1984]
RC683.5.P75C57 1985 612'.1 85-14810

This work is subject to copyright. All rights are reserved, whether the whole or part of the material is concerned, specifically those of translation, reprinting, re-use of illustrations, broadcasting, reproduction by photocopying machine or similar means, and storage in data banks. Under § 54 of the German Copyright Law, where copies are made for other than private use, a fee is payable to "Verwertungsgesellschaft Wort", Munich.

© by Springer-Verlag Berlin Heidelberg 1985
Softcover reprint of the hardcover 1st edition 1985

The use of registered names, trademarks, etc. in this publication does not imply, even in the absence of a specific statement, that such names are exempt from the relevant protective laws and regulations and therefore free for general use.

Product Liability: The publisher can give no guarantee for information about drug dosage and application thereof contained in this book. In every individual case the respective user must check its accuracy by consulting other pharmaceutical literature.

Preface

This volume arose from a discussion meeting entitled "Applied Psychophysiology in Cardiovascular Research: Clinical and Methodological Issues" held in May 1984 in Linz, Federal Republic of Germany. The meeting was organised by the Commission of the European Communities' Concerted Action on Quantification of Parameters for the Study of Breakdown in Human Adaptation, in conjunction with the Department of Internal Medicine, University of Bonn. The Concerted Action is part of the Medical and Public Health Research Programme of the CEC, and its purpose is to encourage multidisciplinary research into the development, refinement and standardisation of parameters used to measure psychosocial, physiological and endocrinological aspects of breakdown. This meeting is one of a series in which detailed consideration is given to specific methodological issues within the field of research linking stress with disease. However, the contributions to this book have been extended and restructured considerably since their presentation at the meeting. Additional chapters have been prepared that address important issues which were not discussed at the meeting itself.

The editors are grateful to the Commission of the European Communities and the University of Bonn for their support, and also acknowledge the financial assistance of Bayer AG, Ciba/Geigy, Hoffman-La Roche, Boehringer Ingelheim and Verla-Pharm. We extend special thanks to Sue Wilson for the care with which she has prepared the manuscript for publication.

March 1985

Andrew Steptoe
Heinz Rüddel
Hermann Neus

Table of Contents

List of Contributors

You will find the addresses at the beginning of the respective contribution

Introduction

The use of psychophysiological techniques in the investigation of cardiovascular diseases has increased rapidly over the past decade. Hypotheses relating emotion and behaviour to the aetiology and prognosis of cardiovascular disorders have led medical and behavioural scientists to explore the cardiovascular and neuroendocrine reactions to psychological stimuli. Much of this research effort has focussed on acute experimental paradigms, although methods of assessing reactivity under field conditions are developing apace.

As work progresses, it has become increasingly apparent that the methodology of acute stress testing requires careful consideration. There are uncertainties and disagreements about what tasks or tests should be presented, what physiological variables should be monitored, and how responses should be analysed. These problems are compounded when psychophysiological test procedures are applied to clinical groups that differ in haemodynamic and medication status. The aim of the present volume is to address some of these issues critically, so that investigators new to the field will benefit from the experience of those who have already encountered methodological and conceptual difficulties. Awareness of these issues has led to parallel efforts in the United States to tease out similar problems (Matthews et al, 1984).

The contents of this book are divided into two sections. Section I concerns the types of psychological stress test used in acute clinical settings, and the bases from which appropriate tasks should be selected. A clinician not familiar with this field might reasonably ask 'why not use the traditional mental stress test of counting backwards aloud?'. This procedure is simple to administer, and has produced useful information concerning cardiovascular reactivity to psychological stress. The contributors to Section I detail why such a test is not ideal for both methodological and conceptual reasons, and outline various alternatives. In Chapter 1, the choice of tasks is considered from a theoretical perspective, from which it is apparent that the type of task selected depends both on the issue being investigated and on the clinical group or pathological process being explored. The influence of the varying demands imposed by different tasks on cardiovascular reactions is detailed in Chapter 2. Factors such as active versus passive coping, control and lack of control, attentional and emotional demands, and situational variables are discussed.

The use of information processing tasks in acute stress paradigms is outlined in Chapter 3. Different types of cognitive and psychomotor task elicit diverse patterns of cardiovascular reaction, but the imposition of steady state mental loads may permit complex regulatory processes such as baroreceptor reflex sensitivity to be assessed. In Chapter 4, it is emphasised that reactions to tasks are most properly evaluated in terms of behavioural performance, and that cardiovascular responses should be seen as adjuncts of such behaviours. The comparison of tasks varying along a single dimension (for instance stimulus presentation time) will greatly facilitate the analysis of concepts such as active behavioural coping.

The varying patterns of haemodynamic response associated with different tasks are detailed in Chapter 5. These reaction patterns are related to the aetiology of hypertension, and it is argued that blood pressure hyperreactivity in borderline and labile hypertensive subjects may be elicited from a narrower range of stimuli than that which provokes hyperreactivity in more severe hypertensives. Hyperreactivity to active coping tasks may be involved in the pathogenetic process, or may be a marker for future hypertension. In Chapter 6, the application of mental stress testing outside the psychophysiology laboratory is described. It is evident that clinically useful data may be collected using quite simple cardiovascular measures in the clinician's office or in field settings, provided that care is taken about baseline assessments and standardisation of task administration. A general point that emerges in all the chapters of Section I is that cardiovascular reactions are determined by a complex interaction between task parameters and individual subject characteristics. The individual differences may be of both a psychological and haemodynamic nature. Consequently, patterns of cardiovascular response cannot be predicted solely from the type of task adminstered in clinical investigations.

<u>Section II</u> of this volume addresses methodological issues in stress test administration and analysis. Chapters 7 and 8 consider the vexed problem of adaptation, and the fact that cardiovascular and neuroendocrine reactions frequently diminish on repeated presentation. In Chapter 8, it is clearly shown that the stability of reactions across occasions varies not only with the nature of the task, but also with the physiological parameter being monitored. The importance of considering the pattern of change within task periods is emphasised in Chapter 7, since valuable information concerning regulatory processes may otherwise be neglected. The relevance of gender and menstrual cycle to tests of women is also detailed.

The question of how to quantify and analyse cardiovascular reactivity is discussed in Chapter 9. The empirical comparison of different change scores and range correction techniques provides a valuable supplement to theoretical arguments concerning types of analysis. Although range-corrected scores seem less stable and robust than simple arithmetical change scores in Chapter 9, they may have important applications in certain clinical populations. It is shown in Chapter 10 that differences in cardiac reactivity among post-infarction patients and other groups are more apparent when data are related to maximal heart rate, and that this range correction also improves the correlations between laboratory and ambulatory measures.

Another problem that arises in the investigation of clinical populations is the effect of medications on psychophysiological responses. Antihypertensive medication is considered in Chapter

11, where it is apparent that testing patients in the few days following withdrawal may lead to serious errors in the quantification of reactivity. The assessment of catecholamine reactions in acute stress paradigms has increased in sophistication over recent years, and current techniques and methodologies are outlined in Chapter 12. As with the study of autonomic parameters, the type of response pattern that is recorded depends to a marked degree on sampling rates and on methods of assay and statistical analysis. The relevance of phasic patterns of heart rate response to clinical issues is discussed in Chapter 13. Although phasic responses are more commonly assessed in relation to non-clinical questions in cardiovascular psychophysiology, they may provide important additional tools for the investigation of parasympathetic influences on the heart. Finally in Chapter 14, methods of analysing fluctuations in cardiovascular activity are outlined, together with an introduction to spectral analytic procedures. These methods too may prove valuable in assessing the relative contribution of the two branches of the autonomic nervous system in clinical psychophysiological studies.

It should be pointed out that in order to focus this book on critical methodological issues, several aspects of clinical cardiovascular psychophysiology have not been discussed. A detailed consideration of the relationship between laboratory and ambulatory measures is beyond the scope of this volume. Ambulatory assessments of cardiovascular and neuroendocrine parameters incorporate an additional series of methodological problems that require more space than can be given here. Secondly, the question of which physiological and endocrine parameters to measure has not been discussed. Although data are presented from a number of different parameters (including blood pressure, cardiac output and its components, peripheral resistance, respiratory variables and catecholamines), the advantages and disadvantages of these measures and their relationship with clinical problems are not described. Finally, we have decided not to elaborate the hypotheses relating these acute reactions with cardiovascular disease. The notion underlying the entire volume is that cardiovascular and neuroendocrine reactions to behavioural stress are related to aetiology and prognosis. However, the extensive animal and human data supporting this proposition cannot be detailed here.

REFERENCE

Matthews KA, Weiss SM, Detre T (1984) Status of and prospects for stress, reactivity and cardiovascular disease: themes from the NHLBI-University of Pittsburgh Working Conference. Behav Med Update 6: 7-9

Section I

The Choice of Psychological Stress Tests

CHAPTER 1

Theoretical Bases for Task Selection in Cardiovascular Psychophysiology

Andrew Steptoe

Department of Psychology, St George's Hospital Medical School, University of London, LONDON SW17 0QS, United Kingdom

SUMMARY

A number of theoretical bases for task selection in acute cardiovascular mental stress testing are considered. They include basing tasks on the demands of everyday life, on the putative psychological characteristics of the disease group, on the metabolic appropriateness of physiological reactions, and on the pathophysiology of the disorder being studied. The merits and limitations of these approaches are outlined. Finally, practical criteria are detailed that stress tests ideally should fulfil.

INTRODUCTION

Research in cardiovascular psychophysiology has numerous aims, many of which are unrelated to clinical issues. However, in the clinical context, stress testing is conducted for two main reasons. The first and most important purpose is the investigation of pathogenic processes, and the role of behaviour and emotion in cardiovascular disorders. Acute stress testing is being used increasingly in studies of the aetiology of essential hypertension, the triggering of ischaemic episodes in patients with angina, prognosis following myocardial infarction and other fields. The second clinical application of stress testing is in the assessment of interventions. Stress testing may provide insight into the psychophysiological mechanisms mediating treatment responses, particularly when behavioural techniques are involved in patient management.

Numerous different tasks are currently employed in clinical stress testing, and the choice of a suitable task for any particular study is far from clear. The aim of the present chapter is to consider the theoretical grounds for task selection. Are there rational reasons for choosing one task over another? In an attempt to resolve this issue, a number of different bases for task selection will be discussed, and their strengths and limitations described.

Historically, the tasks used in clinical stress testing have been selected on a pragmatic basis. A particular stimulus (such as mental arithmetic) might be preferred for its ease of administration, and because the investigators believe it has the correct properties (arousing, emotionally or intellectually challenging etc). Should the task fail to provoke the anticipated pattern of reaction (for example, greater blood pressure increases in hypertensives compared with normotensives), then it might be discarded and replaced with an alternative task. Such an ad hoc approach may have been valuable in the early stages of the development of the discipline, establishing that psychological stimulation could indeed influence blood pressure, coronary blood flow, venous compliance, baroreceptor sensitivity and other processes. However, this method of task selection is atheoretical and essentially pre-scientific. It assumes that stimuli are interchangeable and non-specific, and failures of hypothesis testing cannot be adequately resolved. In order to generate and systematically test notions relating psychological factors to cardiovascular disease, it is necessary to establish adequate theoretical foundations for task selection in stress testing. Four approaches are considered in this chapter.

THE DEMANDS OF ORDINARY LIFE

The first possibility is to choose tasks that reflect the emotional and behavioural demands of everyday life. One might, for example, devisė information processing and psychomotor tasks that call on the cognitive processes that are mobilised during an ordinary working day. This strategy has been influential in ergonometry and studies of the psychophysiology of mental work. The cardiovascular responses associated with different forms of information processing are described in Chapter 3. Tasks have been developed that mirror monotonous working conditions (Lorimer et al 1971), problem solving under time pressure (Schijndel et al 1984) and other common demands. Such an approach has much to recommend it in the clinical field as well, not least because it has ecological validity; tasks selected on this basis have some recognisable relationship with everyday life, instead of being arbitrary abstract inventions of the scientist. A close link with the demands of everyday life is also important as far as pathogenic hypotheses are concerned. Any cardiovascular reaction to behavioural stimulation is only likely to have aetiological significance if similar conditions are met regularly during the course of ordinary life. The probability of a reaction being relevant to the disease process will thus be enhanced if the test calls on resources that are frequently used in occupational or leisure activities. Harshfield et al (1984) have recently reported associations between blood pressure reactions to a mental arithmetic task or video game and fluctuations measured on an ambulatory basis in untreated borderline hypertensives. Reactions to mental arithmetic correlated with blood pressure levels during desk work, while video game reactions predicted blood pressure during other activities.

While this factor must be borne in mind when choosing tasks, it cannot be the sole criterion for selection. The reason is that not all everyday psychological demands provoke potentially pathogenic physiological reactions. A test paradigm might be developed that mirrors the activities of ordinary life precisely, but has absolutely no relevance to clinical disorders. This criterion alone provides no technique for selecting which are the appropriate stimulus and behavioural response dimensions.

PERSONAL CHARACTERISTICS AND THE PSYCHOSOCIAL ENVIRONMENT

A second, radically different approach to task selection is to base the choice on the personality factors or or psychosocial environments that are characteristic of the disease group under investigation. The aim is to devise experimental procedures that mobilise the specific psychological processes thought to be important in the disorder. For example, it has been argued that suppressed hostility is a feature of essential hypertensives. Consequently, cardiovascular reactions have been recorded during assertive role-play from students with and without a parental history of hypertension, and from adult hypertensives (Keane et al 1982; Holroyd and Gorkin 1983). Other methods of studying hypertensives during putatively 'psychologically relevant' challenges include interviews about areas of psychic conflict (Svensson and Theorell 1982), and assessments during tests of social competence (Linden and Feuerstein 1983).

One problem with this approach is that the psychological factors associated with many cardiovascular disorders are far from clear-cut. In the case of hypertension, there are doubts about whether any personality factors or behavioural styles are consistently present; methodological difficulties in patient selection have compromised many studies (Steptoe 1984a). However, the strategy has been successful in other cases, most notably with Type A coronary-prone behaviour. There is general agreement that cardiovascular hyperreactivity is not elicited from Type A individuals in response to all experimental stimuli, and many negative results have been reported (Holmes 1983). However, when the type of experimental stimulus is taken into account, the data are more consistent. Hyperreactivity tends to be observed in Type A people when they are confronted with demanding, problem-solving tasks, or when they are annoyed or harrassed (see Chapter 2). Exaggerated reactions are less common when little effort or challenge is present. In short, the prevailing experimental strategy in this field is to apply stress tests that provoke Type A behaviour (Krantz and Manuck 1984).

It may be that this approach will be extended in the future to explore experimental analogues of particular psychosocial environments that are linked with coronary heart disease. For example, the dimensions of psychosocial work environment identified by Karasek (job demand and job decision latitude) have been associated with risk of myocardial infarction across occupational groups (Alfredsson et al 1982). If these dimensions were operationalised experimentally, studies of psychophysiological reactivity might throw light on the biological processes mediating the increased risk in those with high job demands but low 'discretion' or control in their working lives.

Nevertheless, a danger inherent in this basis for task selection must be considered. As with the first method discussed above, this approach does not specify which physiological parameters are relevant. Consequently, it is not clear whether reactions are causally implicated in the disease process, are parallel but independent effects, or indeed are epiphenomena of no significance whatever. Taking the example of Type A behaviour, a wide variety of different cardiovascular and neuroendocrine parameters have been recorded in psychophysiological studies. When investigations of disease-free adult 'non-student' populations are considered, the most consistent hyperreactivity appears to be present in systolic blood pressure (see Steptoe 1984b). But is systolic pressure reactivity at all relevant to the link between behaviour pattern and heart disease? Perhaps it is a product of some underlying psychobiological process that has no causal significance. There is a tendency to assume that because a physiological response is elicited under psychologically relevant conditions, that the response parameter must be implicated aetiologically. In order to overcome this problem it is necessary to incorporate physiologically based strategies of task selection, in which the psychobiological consequences of the reaction are clear. Two such strategies are outlined below.

METABOLIC APPROPRIATENESS OF PHYSIOLOGICAL REACTIONS

A physiological perspective on reactions of potential clinical significance may take the metabolic appropriateness of the response pattern into account. It can be argued that under most conditions, the cardiovascular adjustments that accompany behavioural responses are congruent with the motor (metabolic) demands on the organism (Cohen and Obrist 1975). If this is the case, the physiological reactions will be similar to those produced at moderate levels of physical exercise, and so are unlikely to be of pathological significance. It is only when cardiovascular reactions are in excess of those required for energy expenditure that a physiological stress response can be said to have occurred. Experimental tasks in which the physiological reactions are out of proportion with motor output should thus be selected. Metabolically inappropriate reactions are thought to be products of the augmented autonomic tone or neuroendocrine output that might mediate links between psychological processes and cardiovascular disease (Obrist 1981).

At one level, this criterion is fulfilled by many experimental tasks. If energy expenditure is measured in terms of the motor response requirement of the task itself, cardiovascular adjustments are frequently excessive. For example, many tasks involve only a verbal response from subjects, while others have key-pressing or similar trivial behavioural responses. It has been shown that talking alone provokes a modest haemodynamic adjustment, and it is interesting that these reactions are greater in hypertensives than normotensives (Friedmann et al 1982). These responses are however of a minor order in comparison with those typically elicited by challenging tasks with verbal responses. Other investigators have shown that supplementary environmental stimulation (eg intermittent pink noise) augments cardiovascular reactions during task performance, even though the task itself is unchanged (Ray et al 1984). Many acute stress tests require no motor response at all during task performance (such as the mental arithmetic tasks described in Chapters 5 and 6), yet substantial haemodynamic and neuroendocrine adjustments are recorded.

However, the criterion of metabolic appropriateness is often defined in a more rigorous fashion, and related to total metabolic activity, and not merely motor response requirement. Although the task itself may require little energy expenditure, subjects engage in other activities (disturbed breathing patterns, contraction of non-involved muscles) that augment motor output. If the total metabolic activity is recorded, cardiovascular adjustments may remain in proportion. It is only when this relationship is violated that, in Obrist's terms, cardiac-somatic uncoupling is deemed to have occurred.

The assessment of this pattern requires measurement of overall energy expenditure during task performance, for example with recording of muscle tension from non-involved muscles or monitoring of oxygen consumption. An important procedure was devised by Blix et al (1974). Initially the normal relationship between oxygen consumption and heart rate was established by monitoring during treadmill exercise. The appropriate heart rate for any degree of oxygen consumption could then be predicted. These researchers subsequently showed that during complex flight operations, the heart rates of helicopter and aircraft pilots were well in excess of the levels expected in terms of oxygen consumption. The 'additional' heart rate recorded under these circumstances represents a metabolically unwarranted cardiovascular adjustment.

This methodology has been extended to laboratory tasks, in an attempt to identify the circumstances under which cardiac-somatic uncoupling takes place. Studies in rats and dogs have demonstrated metabolically inappropriate cardiac reactions during aversive conditioning procedures (Langer et al 1979; Brener et al 1983). Empirical demonstrations that cardiovascular adjustments to experimental tasks in humans are metabolically unwarranted are sparse at present. Gliner et al (1979) showed that the increases in cardiac index during the anticipation of an important public speech were similar to those produced with bicycle ergometry, even though the metabolic demands of the two conditions were quite different. Lovallo et al (in press) compared passive exposure with effortful control of aversive stimulation. Although the active behavioural coping condition produced greater increases in blood pressure, heart rate and indices of myocardial contractility, larger changes in trapezius muscle tension were also recorded, indicating greater muscular effort in this condition. Studies monitoring ventilatory and cardiovascular parameters simultaneously during tasks demanding active behavioural coping suggest that metabolically inappropriate cardiovascular adjustments are produced by only a proportion of subjects (Turner et al 1983; Langer et al, in press). It has been argued that these excessive cardiovascular reactions are mediated through heightened stimulation of the cardiac sympathetic and other ß-adrenergic terminals. Langer et al (in press) found that the additional heart rate provoked during shock avoidance reaction time was blocked with propranolol. Similarly, ß-blockade reversed the excessive systolic blood pressure, heart rate and carotid dp/dt reactions during shock avoidance in the studies reported by Obrist et al (1978).

How relevant are these arguments to acute stress testing in clinical studies? A number of limitations must be considered. Firstly, demonstrations of metabolically unwarranted cardiovascular adjustments have been made with respect to only a few tasks at present. It may be unwise to argue by analogy to other tasks that apparently elicit similar patterns of haemodynamic response. Secondly, the inferences that can be made from pharmacological blockade studies are limited. Compensatory mechanisms come into play during blockade, and these may confuse the response pattern. For example, Bonelli et al (1979) reported that although ß-blockade attenuated the heart rate and cardiac output reactions to mental arithmetic, systolic and diastolic pressure reactions were unchanged. Thirdly, and perhaps most importantly, arguments concerning metabolic appropriateness are difficult to extend beyond the relationship between cardiac parameters (cardiac output and heart rate) and energy expenditure. When other cardiovascular parameters are studied, predictions are less clear. Is it possible to establish whether a blood pressure reaction is metabolically warranted? This cannot be done with the physical exercise methodology utilised for heart rate, since blood pressure reactions are far less consistent. Blood pressure reactions during exercise testing may themselves be metabolically inappropriate. This view is endorsed by studies indicating that exaggerated blood pressure reactions to ergometric exercise predict future hypertension (Wilson and Meyer 1981; Dlin et al 1983). Finally, in many clinical investigations, it is not the overall relationship with energy expenditure that is relevant, but the specific influences on regional haemodynamics. Studies of myocardial ischaemia and coronary spasm are concerned with effects on precisely defined pathophysiological processes, and these may occur whether or not cardiovascular adjustments are in accord with general metabolic demands (Deanfield et al 1984). This suggests that task selection should take account of the pathophysiology of the clinical problem under investigation.

PATHOPHYSIOLOGICAL RELEVANCE OF CARDIOVASCULAR REACTIONS

An alternative basis for choosing tasks in clinical settings is founded on the pathophysiological relevance of reaction patterns. Perhaps tasks should be selected because they provoke haemodynamic or neuroendocrine reaction patterns that are characteristic of the disease under investigation. The reactions may illustrate in the short term and on a small scale the pattern of haemodynamic dysfunction manifest chronically in the disorder itself. This criterion is really the psychophysiological version of a more general strategy in pathophysiology, which is to model experimentally the processes operating in disease. It is the principle that underlies experimental models not only of cardiovascular disease, but also of pathology in other organ systems. One implication of this approach is that the relevance of tasks to disease varies according to the physiological reactions they elicit. The strategy has recently proved valuable in integrating observations on reaction patterns in essential hypertension.

A considerable body of evidence indicates that the haemodynamic processes underlying raised arterial pressure vary with the progression of hypertension. In early or moderate hypertension, elevated blood pressure is commonly sustained by a high cardiac output, in contrast to the raised total peripheral resistance observed in established cases (Lund-Johansen 1980). Although the chain of events is far from clear, heightened sympathetic nervous tone may be important in the initiation of the hypertensive process. Secondary adjustments in renal and vascular mechanisms may subsequently maintain the elevated blood pressure (Guyton 1980; Goldstein 1983). If this is the case, psychological tasks provoking prominent cardiac sympathetic and catecholamine reactions may be the most significant for the analysis of psychophysiological processes in essential hypertension.

It is shown in Chapter 2 that active behavioural coping tasks elicit haemodynamic adjustments sustained by raised cardiac output. McCubbin et al (1983) found that increases in plasma adrenaline levels during performance of visual puzzles correlated positively with systolic blood pressure reactions, and with indices of myocardial contractility. Associations with noradrenaline were also observed. In contrast, the blood pressure increases during passive tasks such as the cold pressor are maintained predominantly by elevations in total peripheral resistance (see Chapter 5). LeBlanc et al (1979) reported that blood pressure reactions to the cold pressor were uncorrelated with plasma adrenaline changes.

The inconsistencies in the literature concerning haemodynamic reactivity in hypertensives and normotensives can be reconciled to using the principle of pathophysiological relevance. Blood pressure hyperreactivity is obtained most consistently in hypertensives when they are exposed to challenging, problem solving tasks. These differences in reactivity are sustained primarily through augmented heart rate and cardiac output reactions (see Chapter 5). Reactions to passive stressors, or undemanding information processing tasks, are often of a similar magnitude in normotensives and hypertensives. Moreover, blood pressure hyperreactivity is seen in response to active challenging tasks in prehypertensive groups as well (see Steptoe 1984b); these data are described in more detail in Chapters 2 and 5. They suggest that hyperreactivity precedes the development of fixed hypertension, and is not a secondary effect of the clinical condition. Whether or not hyperreactivity is significant aetiologically remains to be resolved in prospective investigations.

The selection of tasks on the basis of the pathophysiological relevance of reaction patterns may be extended in the future to other cardiovascular disorders. Already there are indications that the psychophysiological analogues of sudden cardiac death may be very different from those related to hypertension, and implicate other types of task (Verrier et al 1983). Deanfield et al (1984) have demonstrated that mental arithmetic may provoke regional myocardial perfusion abnormalities and ST-segment depression reminiscent of symptomless ischaemic episodes in patients with angina pectoris. Specific experimental models may derive from these types of observation.

SYNTHESIS AND EMPIRICAL CONSIDERATIONS

This discussion of strategies of task selection is not exhaustive, and other bases have been proposed. For example, Dimsdale (1984) suggests the laboratory tasks are compared with profoundly stressful field experiences, and that tasks should be selected if the reactions elicited closely resemble those in the field. Nor are the approaches to task selection outlined in this chapter mutually exclusive. Some tasks fulfil many criteria, duplicating everyday demands and inducing physiological reactions that are metabolically unwarranted and relevant to the pathology of the disorder under investigation. However, it is only by specifying the rational basis of task selection that testable experimental hypotheses can be generated. Otherwise, there is a danger that failures to record the anticipated reactivity differences under varying stimulus conditions or in different clinical groups can be dismissed or excused, and that hypotheses become unfalsifiable.

Another general observation concerning these approaches to stress testing is that they indicate that tasks are not interchangeable. Particular stimuli provoke reactions of significance in the analysis of particular disorders. The task dimensions that determine the pathophysiological response are discussed in more detail elsewhere in this volume.

Task selection cannot be made on the basis of theoretical notions alone, for stimuli might be devised that are quite impractical and unsuitable in clinical settings. A number of empirical or practical criteria must also be considered. These have been summarised elsewhere as follows (see Steptoe 1984c for more details):

(1) It should be possible to administer the test unobtrusively, so that it does not interfere with cardiovascular monitoring.

(2) Cardiovascular responses should be reproducible.

(3) Tests should be largely independent of intellectual status. If task performance depends primarily on intelligence, some people may perform the task more easily, efficiently and rapidly than others, and may therefore produce smaller cardiovascular reactions. Comparisons between clinical groups may then be compromised. This issue is discussed in more detail in Chapter 3.

(4) The test should actively engage the mental and emotional resources of the individual under investigation. It must in some sense be relevant to the participant.

(5) The motor response components of the procedure should be minimal and standardised. If the test involves extensive motor activity, these actions may themselves provoke cardiovascular adjustments irrespective of the psychological challenge. If the task is not paced externally, subjects may produce different cardiovascular reactions purely on the basis of their work rate.

These criteria place limitations on the types of task that can be administered effectively in clinical settings. Some solutions to these problems are discussed elsewhere in the volume. However, it is clear that the selection of tasks in clinical cardiovascular psychophysiology requires careful analysis of the aims of the investigation. The choice of tasks or acute stress tests is not arbitrary, but depends on many theoretical and practical factors, and their relationship to the clinical condition being studied.

REFERENCES

Alfredsson L, Karasek R, Theorell T (1982) Myocardial infarction risk and psychosocial work environment: an analysis of the male Swedish working force. Soc Sci Med 16: 463-467

Blix AS, Stromme SB, Ursin H (1974) Additional heart rate - an indicator of psychological activation. Aerospace Med 45: 1219-1222

Bonelli J, Hörtnagel H, Brüke T, Magometschnigg P, Lochs H and Kaik G (1979) Effect of circulation stress on haemodynamics and plasma catecholamines before and after ß-blockade with propranolol (Inderal) and mepindolol sulphate (Corindolan). Europ J Clin Pharmacol 15: 1-8

Brener J, Phillips K, Sherwood A (1983) Energy expenditure during response-dependent and response-independent food delivery in rats. Psychophysiology 20: 384-392

Cohen DH, Obrist PA (1975) Interactions between behavior and the cardiovascular system. Circulation Res 37: 693-706

Deanfield JE, Shea M, Kensett M, Horlock P, Wilson RA, de Landsheere CM, Selwyn AP (1984) Silent myocardial ischaemia due to mental stress. Lancet ii: 1001-1004

Dimsdale JE (1984) Generalising from laboratory studies to field studies of human stress physiology. Psychosom Med 46: 463-469

Dlin RA, Hanne N, Silverberg DS, Bar-Or O (1983) Follow-up of normotensive men with exaggerated blood pressure response to exercise. Amer Heart J 106: 316-320

Friedman E, Thomas SA, Kulick-Ciuffo D et al (1982) The effects of normal and rapid speech on blood pressure. Psychosom Med 44: 545-553

Gliner JA, Bedi JF, Horvath SM (1979) Somatic and non-somatic influences on the heart: haemodynamic changes. Psychophysiology 16: 358-362

Goldstein DS (1983) Plasma catecholamines and essential hypertension. An analytical review. Hypertension 5: 86-99

Guyton AC (1980) Arterial Pressure and Hypertension. WB Saunders, Philadelphia

Harshfield GA, Pickering TG, Marion RM, Laragh JH (1984) Do laboratory tests predict blood pressure reactivity during a normal day? Psychophysiology 21: 580-581

Holmes DS (1983) An alternative perspective concerning the differential psychophysiological responsivity of persons with Type A and Type B behavior patterns. J Res Personal 17: 40-47

Holroyd KA, Gorkin L (1983) Young adults at risk for hypertension: effects of family history and anger management in determining responses to interpersonal conflict. J Psychosom Res 27: 131-138

Keane RM, Martin HE, Berler ES, Wootton LS, Fleece L, Williams JG (1982) Are hypertensives less assertive? A controlled evaluation. J Consult Clin Psychol 50: 499-508

Krantz DS, Manuck SB (1984) Acute psychophysiological reactivity and risk of cardiovascular disease: a review and methodologic critique. Psychol Bull 96: 435-464

Langer AW, Obrist PA, McCubbin JA (1979) Haemodynamic and metabolic adjustments during exercise and shock avoidance in dogs. Amer J Physiol 5: H225-H230

Langer AW, Stoney CM, McCubbin JA, Hutcheson JS, Charlton JD, Obrist PA (in press) Cardiopulmonary adjustments during exercise and an aversive reaction time task: effect of ß-adrenoceptor blockade. Psychophysiology

Leblanc J, Coté J, Jobin M, Labrie A (1979) Plasma catecholamine and cardiovascular responses to cold and mental activity. J Appl Physiol 47: 1207-1211

Linden W, Feuerstein M (1983) Essential hypertension and social coping behavior: experimental findings. J Human Stress 9(3): 22-31

Lorimer AR, MacFarlane PW, Provan G, Duffy T, Lawrie TDV (1971) Blood pressure and catecholamine response to 'stress' in normotensive and hypertensive subjects. Cardiovascular Res 5: 169-173

Lovallo WR, Wilson MF, Pincomb GA, Edwards GL, Tompkins P, Brackett E (in press) Activation patterns to aversive stimulation in man: passive exposure versus effort to control. Psychophysiology

Lund-Johansen P (1980) State of the art review. Haemodynamics in essential hypertension. Clin Sci 59: 343s-354s

McCubbin JA, Richardson JE, Langer AW, Kizer JS, Obrist PA (1983) Sympathetic neuronal function and left ventricular performance during behavioral stress in humans: the relationship between plasma catecholamines and systolic time intervals. Psychophysiology 20: 102-110

Obrist PA, Gaebelein CT, Teller ES, Langer AW, Grignolo A, Light KC, McCubbin JA (1978) The relationship among heart rate, carotid dp/dt and blood pressure in humans as a function of the type of stress. Psychophysiology 15: 102-115

Obrist PA (1981) Cardiovascular Psychophysiology. Plenum Press, New York

Ray RL, Brady JV, Emurian HH (1984) Cardiovascular effects of noise during complex task performance. Int J Psychophysiology 1: 335-340

Schijndel Mv, De May H, Näring G (1984) Effects of behavioral control and Type A behavior on cardiovascular responses. Psychophysiology 21: 501-509

Steptoe A (1984a) Psychological methods: an overview of clinical applications. In: Cullen J, Siegrist J, Wegmann HM (eds) Breakdown in Human Adaptation to 'Stress', Vol 1. Martinus Nijhoff, The Hague, p208

Steptoe A (1984b) Psychophysiological processes in disease. In: Steptoe A, Mathews A (eds) Health Care and Human Behaviour. Academic Press, London and New York, p77

Steptoe A (1984c) Validation and quantification of mental stress tests, and their application to acute cardiovascular patients. In: Ballieux RE, Fielding JF, L'Abbate A (eds) Breakdown to Human Adaptation to 'Stress', Vol 2. Martinus Nijhoff, The Hague, p1036

Svensson JC, Theorell T (1982) Cardiovascular effects of anxiety induced by interviewing young hypertensive male subjects. J Psychosom Res 26: 359-379

Turner JR, Carroll D, Courtenay H (1983) Cardiac and metabolic responses to 'space invaders': an instance of metabolically-exaggerated cardiac adjustment? Psychophysiology 20: 544-549

Verrier RL, DeSilva RA, Lowne B (1983) Psychological factors in cardiac arrhythmias and sudden death. In: Krantz DS, Baum A, Singer JE (eds) Handbook of Psychology and Health, Vol 3 Cardiovascular Disorders and Behavior. Lawrence Erlbaum Associates, Hillsdale New Jersey, p125

Wilson NV, Meyer BM (1981) Early prediction of hypertension using exercise blood pressure. Prevent Med 10: 62-68

CHAPTER 2

Task Influences on Behaviourally-Elicited Cardiovascular Reactions

Stephen Manuck, David Krantz and Joanna Polefrone

Clinical Psychology Center, 604 Old Engineering Hall, Department of Psychology, University of Pittsburgh, 4015 O'Hara Street, PITTSBURGH, PA 15260, USA
and
Uniformed Services University of the Health Sciences

SUMMARY

Task characteristics strongly influence both the patterning and magnitude of behaviourally-elicited cardiovascular responses. The major stimulus determinants of cardiovascular reactions include: (a) attentional requirements for sensory intake versus sensory rejection; (b) situational demands for active versus passive coping; (c) the nature of emotional or affective states elicited; and (d) associated task attributes such as novelty, difficulty and the presence of performance-contingent incentives. The interest of many clinical investigators in the assessment of psychophysiologic responses follows from recent speculation that a cardiovascular hyperreactivity to stress may be implicated in coronary heart disease (CHD) and/or essential hypertension. In this regard, it is important to examine effects of stimulus variables on the measurement of individual differences in cardiovascular reactivity and on the relationship of such differences to CHD, hypertension and risk for cardiovascular disease. Available data are mixed concerning the stability of idiosyncratic heart rate and blood pressure responses across varying stimulus conditions; some investigators report substantial intertask correlations, whereas others find only moderate, or little, consistency of responses across tasks. Concerning samples of clinical interest, an increased cardiovascular reactivity among hypertensive patients and in persons with a family history of hypertension is observed most frequently during tasks that require subjects to engage in active coping or mental work (sensory rejection). Situational variables are also important in studies of the psychophysiologic correlates of the Type A (coronary-prone) behaviour pattern. Here, the heightened cardiovascular reactivity of Type A individuals appears to be influenced both by the formal nature of the task employed (ie, active versus passive coping, etc) and by the interpersonal and instructional context in which experimental stimuli are presented.

INTRODUCTION

Clinical investigators interested in assessing cardiovascular reactions to behavioural challenge, or 'stress', often assume that all laboratory stimuli evoke the same pattern of haemodynamic response. In turn, this assumption frequently leads to selection of a single experimental task, or challenge, to be used in studies of behaviourally-elicited cardiovascular reactivity. Yet psychophysiologists have long noted that certain stimulus attributes give rise to characteristically different types of cardiovascular response, and that other task variables influence appreciably the magnitude of observed responses. The purpose of the present chapter is to identify stimulus parameters thought to be important in eliciting significant cardiovascular reactions and to examine what relevance these variables may have for an understanding of individual differences in cardiovascular reactivity. A related purpose is to explore the influence of task parameters on the cardiovascular responses of relevant clinical samples (eg, hypertensives) and of persons at heightened risk for cardiovascular pathology due to behavioural characteristics (eg, Type A behaviour pattern) or family history of heart disease.

STIMULUS DETERMINANTS OF CARDIOVASCULAR RESPONSE

Sensory intake versus sensory rejection

Many years ago, the Laceys reported that both acceleratory and deceleratory heart rate responses could be elicited in laboratory experiments, depending upon the types of attentional demands placed on subjects (Lacey et al 1963). They noted that heart rate decreased while subjects performed tasks requiring vigilance, or attentiveness to an external stimulus (sensory intake), but increased during tasks of a cognitive, or 'mental' nature; the latter were situations which necessitated concentration and, hence, rejection of extraneous sensory information (sensory rejection). Obrist et al (1970) subsequently demonstrated that these different cardiac responses often reflect concomitant changes in somatomotor activity - somatic inhibition occurring in conjunction with sensory intake and heart rate deceleration, and somatic activation accompanying sensory rejection and heart rate acceleration. Other investigators have also examined the influence of sensory intake and sensory rejection on skeletal muscle haemodynamics. In such studies, it is frequently reported that forearm blood flow increases and forearm vascular resistance decreases during sensory rejection tasks, whereas forearm blood flow is decreased and forearm vascular resistance increased during sensory intake (ie, an active vasoconstriction) (Williams, in press).

In summary, the typical response to sensory rejection (or mental work) is tachycardia and a redistribution of blood flow from skin and viscera to skeletal muscle. This pattern may be accompanied by some increase in somatomotor activity. In contrast, sensory intake tasks are associated with decreased motor activity, an unchanged or perhaps lowered cardiac output, decreased heart rate, and vasoconstriction in skin, viscera and skeletal muscle (possibly due to active vasoconstriction).

A few additional points should be noted in relation to this stimulus dimension. First, interpretation of cardiovascular responses in terms of sensory intake and rejection per se is arguable in many studies, since tasks employed in this literature (reaction time tests, serial subtraction, etc) frequently differ in characteristics other than attentional demands (such as motor versus verbal response requirements). Secondly, sensory intake and sensory rejection tasks do not usually elicit opposing cardiovascular reactions of equal magnitude and duration. For example, the heart rate response to mental arithmetic may reach 15 or 20 beats per minute and persist over five or ten minutes. In contrast, heart rate reactions during sensory intake tasks tend to be quite small - perhaps three to four beats per minute - and are usually recorded as phasic phenomena lasting seconds, rather than minutes.

Active Versus Passive Coping

Recently, Obrist (1976) described a second task dimension influencing cardiovascular responses to behavioural stimuli. It was found that experimental subjects exposed to a shock avoidance procedure exhibited large increases in heart rate, heart contractility and systolic blood pressure (Obrist et al 1978). Since these responses could be blocked by infusion of propranolol, they also appeared to be largely beta-adrenergic in origin. Interestingly, these same subjects did not experience comparable reactions to two other tasks that were either aversive (the cold pressor test) or arousing (a pornographic film). Obrist noted that a salient feature of avoidance tasks is the availability of a behavioural response permitting subjects some control over the occurrence of shock. Conversely, in the cold pressor test, subjects have no opportunity to escape or avoid aversive stimulation short of discontinuing their participation in the task. Obrist therefore proposed that the stimulus attribute responsible for eliciting large, sympathetically-mediated myocardial responses (such as those observed during shock avoidance) is the availability of instrumental control over aversive events.

Light and Obrist (1980) later demonstrated the influence of this task variable more persuasively, using a better controlled experimental design. Subjects were all exposed to the same laboratory stimuli (including shock), but differed in whether or not an avoidance response could be made. Subjects having potential control over shock exhibited the greatest increases in heart rate, systolic blood pressure, and indirect indices of myocardial contractility. In a related experiment, Manuck et al (1978) observed that even the perception of control over an anticipated aversive event (ie, without actual control) significantly increased systolic pressor responses.

Similar response patterns were subsequently seen while individuals performed an appetitive reaction time task in which reward, rather than avoidance, was contingent on behavioural performance (Light and Obrist 1983). Such findings led Obrist (1981) to propose a more general distinction between active and passive coping. By this formulation, active coping, which refers to an individual's attempts to exert personal control over either positive or negative environmental events, leads to significant sympathetic myocardial responses. In situations that offer little or no opportunity to exercise instrumental control, or where coping behaviours (if any) are necessarily passive in nature (ie, passive coping), such beta-adrenergic reactions are much attenuated. Since this dimension is not restricted to experimental procedures involving noxious stimulation, the notion of active and passive coping denotes a relatively broad construct encompassing many of the non-aversive activities in which one is motivated to exhibit competent and effective performance. We would also note that such activities subsume many tasks that could otherwise be labelled sensory rejection or 'mental work', as defined in the previous section.

Obrist has emphasised the 'active coping' end of this dimension, and, in particular, its beta-adrenergic influence on myocardial performance. In large measure, this is due to a more general interest of his research group in the autonomic and haemodynamic mechanisms underlying hyperkinetic borderline hypertension. Yet, the responses seen during so-called passive coping tasks are also of interest. Despite their relatively minor effects on heart rate and heart contractility, passive stimulus conditions do elicit a significant vascular response, which is reflected in a marked elevation in diastolic blood pressure. It is also noteworthy that after infusion of propranolol a similar diastolic response is observed among subjects exposed to shock avoidance (Obrist et al 1978). This suggests that the smaller diastolic reaction to 'active coping' conditions in autonomically intact subjects is attributable to vasodilation in skeletal muscle, which effectively offsets concomitant vasoconstriction due to alpha adrenergic stimulation in other tissues.

Affective States

The particular affective or emotional reaction elicited by an experimental task may also influence cardiovascular responses. Some of the earliest studies in experimental psychophysiology sought to demonstrate differentiation of heart rate and blood pressure reactions (as well as other autonomic responses) during laboratory procedures designed specifically to induce emotions such as fear and anger (see Sternbach 1966). More recent investigations have shown that cardiovascular responses differ across a wide range of negative emotions (eg, fear, anger, sadness, disgust) and between positive and negative emotions generally (Ekman et al 1983). Though this is speculative, Ekman et al (1983) hypothesise that such emotion-specific autonomic activity is associated with the patterning of facial muscle contractions characteristic of each emotion, and that these relationships are mediated by feedback from the facial muscles themselves or by 'direct connection between the cortex and hypothalamus'.

The foregoing studies involve rather dramatic experimental manipulations (Ax 1953) or precise quantification of the facial expressions. Studies in which individual differences in emotion or affective response are documented with questionnaires and personality inventories yield a less consistent literature. It has been suggested that lack of concordance between self-report indices and physiological reactions may be due partly to the psychometric deficiencies of 'affect' scales typically employed in laboratory studies. They may be especially insensitive to defensiveness as a source of nonveridical self-report. Thus while emotion is related to the magnitude of cardiovascular responses elicited in laboratory experiments, this relationship may be obscured when relevant affective experiences of the individual are assessed only through self-report instruments.

Other Task Variables

In addition to the major stimulus categories relating to attentional and coping processes, and to emotionality, several additional task variables may be mentioned briefly. These variables have not been studied as fully as those described in the preceding sections; nevertheless, such factors as task difficulty, performance incentives and stimulus novelty do influence at least the magnitude of cardiovascular responses. With respect to novelty, the tonic heart rate elevations observed during shock avoidance tasks and in anticipation of shock appear to be greatest when subjects are provided with no prior exposure to the aversive stimulus (Elliott, 1966; Light and Obrist, 1980). Cardiovascular reactions recorded on the first presentation of an experimental task (a novel experience) are generally greater than those seen when the same task is administered again on subsequent occasions (see Chapters 7 and 8). Inclusion of performance-contingent incentives leads to increased heart rate and blood pressure elevations, which in some cases may be quite appreciable (Manuck and Garland 1979). Whether subjects' cardiovascular reactions also vary in proportion to the value of the incentive applied, however, has not been examined directly.

Effects of differences in task difficulty on cardiovascular responsivity are somewhat more complicated. For example, Manuck et al (1978) reported that increasing the difficulty of an experimental task involving concept forming abilities resulted in heightened systolic pressor response, but only in subjects who believed that performing well would prevent their exposure to a painful auditory stimulus. When subjects believed that exposure to the aversive stimulus was random and independent of performance, their blood pressure responses did not differ from those seen in subjects who were allowed to perform a much easier version of the task.

In another investigation, Obrist et al (1978) assigned three groups to an experimental procedure in which avoidance of electric shocks was contingent on level of performance in a reaction time task; importantly, the criterion for successful avoidance varied in increments across the three groups, which were labelled 'easy', 'difficult' and 'impossible'. Heart rate and systolic blood pressure elevations during the avoidance task were most pronounced among subjects required to perform the difficult reaction time task. Responses in this condition were also significantly greater than those seen when the task was more easily performed as well as in the 'impossible' condition, where subjects were effectively helpless to avoid shock. That similar effects are found when 'difficulty' is manipulated during appetitive, rather than avoidance, tasks suggests also that this relationship is not restricted to situations involving threat of aversive stimulation (Elliott 1969; Light et al 1983). Hence, while increased task difficulty potentiates cardiovascular responses to laboratory challenges, eliciting such effects may require the presence of extrinsic incentives for good performance and the use of levels of difficulty that do not exceed individuals' actual abilities.

TASK CHARACTERISTICS AND INDIVIDUAL DIFFERENCES

It is clear that behavioural stimuli often elicit significant cardiovascular reactions, and that these responses vary with differences in task characteristics. Yet the magnitude of heart rate and blood pressure responses typically recorded in laboratory investigations also varies widely among individuals. It is these individual differences that have recently attracted the interest of many clinicians. This interest arises from a variety of indirect, but suggestive, sources of evidence that suggest that excessive reactivity to stress may play a role in the development and/or clinical expression of certain cardiovascular disorders, including CHD and essential hypertension (Krantz and Manuck, 1984). Hypertensive and CHD patients have been found to exhibit significantly increased heart rate and blood pressure responses when exposed to common laboratory challenges, in comparisons with both patient and nonpatient control samples. Heightened cardiovascular responses have also been observed among normotensive individuals with a family history of hypertension and in persons thought to be behaviourally predisposed to CHD (ie, so-called Type As).

Three assumptions shared by many investigators are that individual differences in behaviourally-induced cardiovascular reactivity show stability over time, correlate highly across a variety of different experimental tasks and stimuli, and generalise from laboratory assessments to field (or natural) settings. The assumptions concerning temporal stability and generalisation are discussed in more detail in Chapters 6, 8 and 9. More directly relevant to the present discussion is the reproducibility of individual differences in cardiovascular reactivity under varying task conditions. Insofar as task dimensions such as sensory intake versus sensory rejection, or active and passive coping, predict differential patterning of cardiovascular reactions, one might expect idiosyncratic heart rate and blood pressure responses to show little reproducibility when assessed under different stimulus conditions. It is also possible, however, that individual differences override the influence of stimulus characteristics in determining the magnitude of observed responses. This issue may be considered empirically in relation to two relatively specific questions: do individuals who differ in their cardiovascular responsivity to one experimental task or challenge show similar differences when exposed to one or more other laboratory stimuli? To what extent is the heightened cardiovascular responsivity seen frequently in samples of clinical interest (eg, hypertensives, Type A individuals) independent of the attributes of eliciting stimuli? Each of these questions is discussed briefly in the remaining sections of this chapter.

Intertask Stability of Individual Differences

Bunnell (1982) demonstrated recently that subjects who differ in the magnitude of their heart rate responses during sensory rejection exhibit parallel, though somewhat smaller, differences when performing sensory intake tasks. Two subject groups (labelled Reactors and Nonreactors) were identified based on their heart rate reactions to mental arithmetic. Reactors showed a mean heart rate acceleration of +17.5 beats per minute and Nonreactors a mean change of -0.8 beats per minute. The corresponding heart rates of Reactors and Nonreactors during an auditory reaction time task (sensory intake) averaged +2.3 and -4.8 beats per minute, respectively ($p<0.01$). Hence, subjects who failed to show an increased heart rate during mental arithmetic exhibited the expected cardiac deceleration during sensory intake. On the other hand, the most reactive subjects under mental arithmetic showed a slight increase in heart rate while performing the reaction time task. Interestingly, the Laceys estimate that about 25% of individuals do not show deceleration of heart rate when exposed to intake tasks (Lacey and Lacey 1974). Bunnell's findings suggest that these anomolous individuals may, in fact, be those who exhibit the largest heart rate acceleration when engaged in sensory rejection tasks.

If individual differences in heart rate reactivity observed during a sensory rejection task predict responses recorded during sensory intake tasks, is the same true of responses to active and passive

Table 1. Mean heart rate (in beats per minute) during relaxation baseline, and mean heart rate changes during three laboratory tasks, in four groups (n = 14) ranked by level of heart rate reactivity under shock avoidance (adapted from Obrist 1981)

	Heart rate reactivity			
	most reactive			least reactive
	I	II	II	IV
Relaxation baseline	63	65	68	67
Change from relaxation baseline during:				
Shock avoidance	+57	+38	+25	+9
Cold pressor	+33	+27	+19	+13
Pornographic film	+23	+15	+12	+5

coping tasks? Table 1 (adapted from Obrist 1981) lists mean baseline heart rates and heart rate changes during three tasks in groups of college-aged males. The four groups represent quartiles of the distribution of heart rate changes observed during shock avoidance, and are ranked from most to least reactive. Note that while baseline heart rates of the four groups are comparable, mean heart rate responses during shock avoidance (active coping) vary considerably, from +9 to +57 beats per minute. The rows labelled Cold pressor and Pornographic film list corresponding heart rate changes during the two passive coping tasks. While the magnitude of group differences under these conditions is much attenuated, rank order relationships between the four groups are nonetheless retained. In addition, the correlations between heart rate changes among the three experimental tasks (calculated over all subjects) are significant (r ranging from +0.53 to +0.58). Correlations between systolic pressor responses recorded during the same tasks are also significant and of similar magnitude (r = +0.49 to +0.75). Thus, persons who show different heart rate and systolic blood pressure reactions during at least one active coping task (shock avoidance) also tend to differ when exposed to painful or arousing stimuli over which they have little or no instrumental control.

It is important to note that heart rate and blood pressure responsivity was assessed relative to a 'relaxation' baseline recorded under maximally resting conditions and on a day following the administration of experimental stressors. Calculation of intertask correlations based on heart rate or systolic pressure changes from a second baseline, obtained just prior to the presentation of tasks, reveals no significant relationships among subjects' responses to the three task conditions. Pre-task values were apparently themselves elevated among more reactive subjects, resulting in an underestimation of the heart rate and blood pressure reactions of hyperresponsive individuals. This finding is of potential significance since 'reactivity' is often expressed as a simple change score, reflecting the absolute difference between measurements taken during and immediately preceding performance of experimental tasks.

Other investigators have occasionally reported correlations between measures of heart rate and blood pressure reactivity, as observed in the same subjects under different task conditions. In these studies, however, experimental stimuli were not selected for their relevance to major task dimensions such as sensory intake versus sensory rejection, or active and passive coping. Nevertheless, Rüddel et al (1983) reported that in a sample of male students, systolic pressor responses during mental arithmetic covaried significantly (r = 0.50) with systolic changes seen on administration of the Structured Interview for assessment of Type A behaviour pattern. Manuck and Garland (1980) found that among young adult subjects asked to perform two cognitive tasks - mental arithmetic and a concept forming test - the intertask correlation was very high for heart rate responses (r = 0.86), moderate for systolic pressure reactions (r = 0.50), and only marginal for diastolic responses (r = 0.23 NS). In a third data set, reported by Hill and Krantz (1983), medical school students were presented three tasks: a reaction time task, a medical quiz, and the Type A Structured Interview. Intertask correlations here were significant for all response measures - heart rate, systolic and diastolic blood pressure - and ranged from 0.53 to 0.81.

The foregoing studies, and others not described here, present a mixed picture regarding the influence of task variables on stability of individual differences. In some instances, the distributions of heart rate or blood pressure reactions to one experimental task appear to be largely retained when subjects are presented a second or third task to perform - even when these tasks are conceptually quite dissimilar or reflect differences in the major stimulus variables that determine overall patterning of cardiovascular response. However, other investigators find little consistency across tasks, or observe consistency on some response parameters and not others. Among studies in which subjects' cardiovascular reactions are correlated significantly across task conditions, moreover, the strength of these associations tends only to be of a moderate order of magnitude.

Discrepancies in this literature may reflect a host of methodologic differences among the several studies cited (Krantz and Manuck 1984). Still, available data do not support speculation that there exists an underlying dimension of individual differences in cardiovascular reactivity, the expression of which is entirely independent of stimulus characteristics. From an assessment standpoint, then, it is unlikely that use of any one experimental challenge will allow an adequate evaluation of the psychophysiologic response characteristics of individuals. Such an evaluation may instead require the construction of a battery of test stimuli, sampling a variety of task variables known to affect cardiovascular responses. In fact, the most interesting dimension of individual differences may not be the distribution of cardiovascular responses to a single behavioural challenge, but the extent to which intertask consistency is exhibited by each subject. Some persons, for example, may show exaggerated pressor reactions under all task conditions, while others may reveal a more variable pattern of responses; in turn, it may be the stereotypically hyperresponsive individual who is at greatest risk for any clinical sequelae that may follow from a heightened cardiovascular reactivity to stress.

Cardiovascular Reactivity in Hypertension and Coronary Heart Disease

Aside from the stability of cardiovascular reactions under varying stimulus conditions, it may also be asked whether task variables influence cardiovascular reactivity in clinical populations. As noted previously, results of several investigations indicate that hypertensive individuals show larger blood pressure and/or heart rate responses to laboratory stressors than normotensive controls. Such differences have been seen, for example, during experimentally-elicited emotions of fear and anger (Schachter 1957), during conflict-laden interviews (McKegney and Williams 1967), on performance of frustrating cognitive tasks (Nestel 1969), and, somewhat inconsistently, during the cold pressor test (see Chapter 5). Yet, importantly, some researchers have also failed to find significant associations (Sullivan et al 1981).

Few investigators have directly examined the possibility that stimulus characteristics mediate relationships between hypertensive status and cardiovascular responsivity. Nevertheless, what evidence exists suggests that increased reactivity among hypertensive patients does depend, in part, on situational factors. Fredrikson et al (in press) report, for instance, that when compared to normotensive controls, hypertensives exhibited heightened pressor responses in the initial phase of a sensory rejection (mental work) task - mental arithmetic - but not during a sensory intake task involving identification of alphabetic letters. In a recent investigation by Steptoe et al (1984), hypertensive individuals showed greater pressor responses than normotensives when engaged in tasks involving forms of active coping (such as video games and the Stroop colour-word interference task), but not during passive exposure to a distressing film. Drummond (1983) reports, too, that when compared with normotensive controls, hypertensive subjects exhibited greater elevations in systolic and diastolic blood pressure under mental arithmetic, but not during the cold pressor test. The passive aspect of cold immersion may also account for equivocal results among the many studies in which only the cold pressor test has been administered (Boyer et al 1960).

A related literature concerns the psychophysiologic response characteristics of normotensive persons with and without a family history of essential hypertension. Like hypertensive patients, offspring of hypertensives also tend to exhibit heightened blood pressure and heart rate responses to laboratory stressors (eg, Falkner et al 1979). The experimental stimuli employed in these studies have varied considerably, from the cold pressor test and serial subtraction to shock avoidance. In addition, reports by Jorgensen and Houston (1981) and by Manuck and Proietti (1982), who exposed subjects to a number of different stressors, suggest that the heightened cardiovascular reactivity of persons at familial risk for hypertension generalises across varying stimulus conditions. This generality may be more apparent than real, however, as the multiple tasks employed in both of these studies are conceptually quite similar - mainly, active coping and mental work tasks such as mental arithmetic, concept formation, 'digits backwards' from the Wechsler scales, and shock avoidance. Hastrup et al (1982), on the other hand, have contrasted the cardiovascular responses of offspring of hypertensive and normotensive parents to both shock avoidance and cold pressor tests (ie, active and passive coping tasks). As in the Steptoe et al (1984) and Drummond (1983) studies of hypertensive subjects, Hastrup et al (1982) observed significant group differences in heart rate and systolic pressor responses during the active task - avoidance - but only marginal differences in reactions to cold immersion.

Acute psychophysiologic reactivity has been implicated in coronary heart disease as well as hypertension. However, unlike hypertension, there are presently few studies contrasting the cardiovascular responses of CHD patients and controls. Most of the case control studies reported to date have shown significantly increased blood pressure responses to laboratory stressors in patients with histories of angina or previous myocardial infarction. Yet, interpretation of results of several of these investigations is hampered by methodologic faults, such as failure to control for the medication status of patients and the presence of other chronic disorders. In relation to task influences, moreover, available studies are too few to make meaningful contrasts, nor has any investigator yet compared CHD patients and controls on tasks selected specifically for their different stimulus characteristics.

On the other hand, there is now a considerable literature concerning the cardiovascular response characteristics of persons believed to be at increased risk for CHD due to their behavioural characteristics - namely, Type A individuals. As described by cardiologists Friedman and Rosenman, Type A (coronary-prone) persons are characterised by an intense, hard-driving competitiveness, a chronic sense of time urgency, and easily-evoked hostility; the absence of these characteristics denotes a more placid, Type B (noncoronary-prone) pattern. About fifty studies have now been reported contrasting the psychophysiologic responses to Type A and Type B individuals. These investigations involve a wide range of psychological and physical stressors (eg, cognitive and psychomotor tasks, the cold pressor test, interpersonal challenges, isometric handgrip) and have employed a variety of subject populations (eg, children, college students, middle-aged adults, CHD patients). Despite several failures of replication, in the majority of published investigations subjects identified as Type As have exhibited larger increases in blood pressure, heart rate, catecholamines and/or cortisol, relative to Type Bs, when exposed to appropriately challenging or stressful laboratory tasks (Wright et al, in press). The strength of these associations depends partly on the method used to assess Type A behaviour (Structured Interview versus questionnaires), and in several studies, differences in the cardiovascular reactions of Type A and Type B subjects emerge primarily among persons having a family history of cardiovascular disease (eg, Williams et al 1982). In addition to these variables, task characteristics are also critical in eliciting a heightened responsivity among Type A individuals.

While the literature on psychophysiologic correlates of Type A behaviour is too extensive to review adequately here, several representative studies relevant to task influences may be described briefly. Dembroski et al (1979) found no relationship between Type A behaviour and cardiovascular responses to an ordinary cold pressor test. Yet when this task was presented under more challenging instructions emphasising its difficulty and the need to show persistence, Type As did exhibit greater heart rate and pressor responses than their Type B counterparts. Interestingly, this finding suggests that the prototypic 'passive coping' task described by Obrist (1976) - the cold pressor test - can take on properties of an 'active coping' task through appropriate manipulation of instructional sets.

Williams et al (1982) compared the cardiovascular and neuroendocrine responses to Type A and Type B subjects to tasks requiring either sensory intake or sensory rejection. During the sensory rejection task (mental arithmetic), Type As exhibited greater changes in forearm blood flow, forearm vascular resistance, and plasma norepinephrine, epinephrine and cortisol, compared to Type Bs. In contrast, neither the haemodynamic nor plasma catecholamine responses differed between Type A and Type B subjects during the sensory intake task (reaction time).

Significant positive correlations were reported by Manuck and Garland (1979) between subjects' scores on a Type A questionnaire (Jenkins Activity Survey) and heart rate, systolic blood pressure and pulse pressure reactions to a frustrating cognitive task. When the same task was presented with a monetary incentive for good performance, though, correlations between Type A behaviour and cardiovascular responses were not significant. The authors suggest that the inclusion of an extraneous reward motivated Type Bs to work as effortfully, during the experiment, as Type As perform (by disposition) in most all work-related situations. This interpretation was consistent with measures of task performance, which showed that Type Bs solved as many test problems correctly as their Type A counterparts only in the incentive condition.

An additional way in which subjects' task appraisals are of importance is illustrated by MacDougall, Dembroski and Krantz (1981). These investigators observed that laboratory stimuli which had previously elicited differences in the cardiovascular responsivity of Type A and B males failed to produce similar results when applied to a sample of undergraduate women. When females were required to interact verbally with a female confederate of the investigator, however, Type A women did show greater heart rate and systolic blood pressure responses than Type Bs. This finding suggests that in certain subject groups, such as young adult women, direct interpersonal challenge may provide a more effective stimulus for eliciting A-B differences in psychophysiologic reactivity.

Finally, recent investigations suggest that among the various Type A attributes, anger or high 'potential for hostility' may be especially important in the development of coronary disease (Matthews et al 1977). Exploring the role of anger in the psychophysiologic differentiation of Type A and B individuals, Glass and co-workers (1980) instructed subjects to perform a competitive video game task against an experimenter's confederate. The confederate, in turn, made either harassing and disparaging remarks toward the subject or interacted in a more neutral fashion. Results revealed that, during the task, Type A subjects showed significantly greater increases in heart rate, systolic blood pressure and plasma epinephrine measurements than Type Bs, but only in the harassment condition. These findings indicate that in competitive situations, instigation of anger is a potent stimulus for sympathetic nervous hyperresponsivity in persons possessing Type A characteristics.

From the preceding examples, it is clear that stimulus factors modulate, to a great extent, the influences of Type A behaviour on cardiovascular responses to laboratory stressors. Particularly important in eliciting Type A-reactivity associations are the use of sensory rejection (mental work) and active coping tasks (Williams et al 1982), interpersonal challenges (MacDougall et al 1981), harassment in conjunction with competition (Glass et al 1980), and instructional sets leading to an increased task involvement (Dembroski et al 1979). That situational variables figure prominently in these studies is not surprising, since the Type A construct is not defined as a personality trait, but as a set of behavioural propensities that are exhibited in interaction with the demands of the social and work environment (Matthews 1982). However, it must still be acknowledged that there are inconsistencies in this literature (see Wright et al, in press, for a comprehensive review). Some investigators have failed to observe significant relationships even when employing stimuli similar to those described above (Light and Obrist 1983).

REFERENCES

Ax AF (1953) The physiological differentiation between fear and anger in humans. Psychosom Med 15: 433-442

Boyer JT, Fraser JRE, Doyle AD (1960) The hemodynamic effects of cold immersion. Clin Sci 19: 539-550

Bunnell DE (1982) Autonomic myocardial influences as a factor determining inter-task consistency of heart rate reactivity. Psychophysiology 19: 442-448

Dembroski TM, MacDougall JM, Herd JA, Shields JL (1979) Effect of level of challenge on pressor and heart rate responses in Type A and B subjects. J Appl Soc Psychol 9: 209-228

Drummond PD (1983) Cardiovascular reactivity in mild hypertension. J Psychosom Res 27: 291-297

Ekman P, Levenson RW, Friesen WV (1983) Autonomic nervous system activity distinguishes among emotions. Science 221: 1208-2120

Elliott R (1966) Effects of uncertainty about the nature and advent of a noxious stimuli (shock) upon heart rate. J Pers Soc Psychol 3: 353-356

Elliott R (1969) Tonic heart rate: experiments on the effects of collative variables lead to a hypothesis about its motivational significance. J Pers Soc Psychol 12: 211-228

Falkner B, Onesti G, Angelakos ET, Fernandes M, Langman C (1979) Cardiovascular response to mental stress in normal adolescents with hypertensive parents. Hypertension 1: 23-30

Fredrickson M, Dimberg U, Frisk-Holmberg M, Ström G (in press) Arterial blood pressure and general sympathetic activation in essential hypertension during stimulation. Acta Med Scand

Glass DC, Krakoff LR, Contrada R et al (1980) Effect of harassment and competition upon cardiovascular and catecholamine responses in Type A and B individuals. Psychophysiology 17: 453-463

Hastrup JL, Light KC, Obrist PA (1982) Parental hypertension and cardiovascular response to stress in healthy young adults. Psychophysiology 19: 615-622

Hill DR, Krantz DS (1983) Type A behavior and cardiovascular response in military medical students as a function of academic stress. Paper presented at meeting of the Eastern Psychological Association, Philadelphia, April

Jorgensen RS, Houston BK (1981) Family history of hypertension, gender, and cardiovascular reactivity and stereotypy during stress. J Behav Med 4: 175-189

Krantz DS, Manuck SB (1984) Acute psychophysiologic reactivity and risk of cardiovascular disease: a review and methodologic critique. Psychol Bull 96: 435-464
Lacey BC, Lacey JI (1974) Studies of heart rate and other bodily processes in sensorimotor behaviour. In: Obrist PA, Black AH, Brener J and DiCara LV (eds) Cardiovascular Psychophysiology. Aldine, Chicago, p538
Lacey JI, Kagan J, Lacey BC (1963) The visceral level: situational determinants and behavioral correlates of autonomic response patterns. In: Knapp PH (ed) Expression of Emotions in Man. International University Press, New York, p161
Light KC, Obrist PA (1980) Cardiovascular response to stress: effects of opportunity to avoid, shock experience and performance feedback. Psychophysiology 17: 243-252
Light KC, Obrist PA (1983) Task difficulty, heart rate reactivity, and cardiovascular responses to an appetitive reaction time task. Psychophysiology 20: 301-312
McDougall JM, Dembroski TM, Krantz DS (1981) Effects of types of challenge on pressor and heart rate responses in Type A and B women. Psychophysiology 18: 1-9
McKegney FP, Williams RB (1967) Psychological aspects of hypertension: II. The differential influence of interview variables on blood pressure. Amer J Psychiat 123: 1539-1545
Manuck SB, Harvey AE, Lechleiter SL, Neal KS (1978) Effects of coping on blood pressure responses to threat of aversive stimulation. Psychophysiology 15: 544-549
Manuck SB, Garland FN (1979) Coronary-prone behavior pattern, task incentive and cardiovascular response. Psychophysiology 16: 136-142
Manuck SB, Garland FN (1980) Stability in individual differences in cardiovascular reactivity: a thirteen-month follow-up. Physiol Behav 21: 621-624
Manuck SB, Proietti J (1982) Parental hypertension and cardiovascular response to cognitive and isometric challenge. Psychophysiology 19: 481-489
Matthews KA (1982) Psychological perspectives on the Type A behavior pattern. Psychol Bull 91: 292-323
Matthews KA, Glass DC, Rosenman RH, Bortner RW (1977) Competitive drive, pattern A, and coronary heart disease: a further analysis of some data from the Western Collaborative Group Study. J Chron Dis 30: 489-498
Nestel PJ (1969) Blood pressure and catecholamine excretion after mental stress in labile hypertension. Lancet I: 692-694
Obrist PA (1976) The cardiovascular-behavioral interaction - as it appears today. Psychophysiology 13: 95-107
Obrist PA (1981) Cardiovascular Psychophysiology: a perspective. Plenum Press, New York
Obrist PA, Webb RA, Sutterer JR, Howard JL (1970) The cardiac-somatic relationship: some reformulations. Psychophysiology 6: 569-587
Obrist PA, Gaebelein CJ, Teller ES, Langer AW, Grignolo A, Light KC, McCubbin JA (1978) The relationship among heart rate, carotid dp/dt and blood pressure in humans as a function of the type of stress. Psychophysiology 15: 102-115
Rüddel H, Gogolin E, Friedrich G, Neus H, Schulte W (1983) Coronary-prone behavior and blood pressure reactivity in laboratory and life stress. In: Dembroski TM, Schmidt TH and Blümchen G (eds) Biobehavioral Bases of Coronary Heart Disease. Karger, Basel, p185
Schachter J (1957) Pain, fear, and anger in hypertensives and normotensives. Psychosom Med 19: 17-29

Steptoe A, Melville D, Ross A (1984) Behavioral response demands, cardiovascular reactivity and essential hypertension. Psychosom Med 46: 33-48

Sternbach RA (1966) Principles of Psychophysiology. Academic Press, New York

Sullivan P, Schoentgen S, De Quattro V, Procci W, Levine D, van der Meulen J, Bornheimer J (1981) Anxiety, anger and neurogenic tone at rest and in stress in patients with primary hypertension. Hypertension 3: II-119-II-123

Williams RB (in press) Neuroendocrine response patterns and stress: biobehavioral mechanisms of disease. In: Williams RB (ed) Perspectives on Behavioral Medicine, Vol 2, Neuroendocrine Control and Behavior. Academic Press, New York

Williams RB, Lane JD, Kuhn CM, Melosh W, White AR, Schanberg SM (1982) Type A behavior and elevated physiological and neuroendocrine responses to cognitive tasks. Science 218: 483-485

Wright RA, Contrada RJ, Glass DC (in press) Psychophysiologic correlates of Type A behavior. In: Katkin ES, Manuck SB (eds) Advances in Behavioral Medicine. JAI, Greenwich CT

CHAPTER 3

Mental Tasks as Stressors

G Mulder, LJM Mulder and JBP Veldman

Instituut voor Experimentele Psychologie der Rijksuniversiteit te Groningen, Biologisch Centrum, Kerklaan 30, 9751 NN HAREN (Gn), The Netherlands

SUMMARY

The cardiovascular reactions to mental tasks may reflect the way in which the subject copes with the demands of the task. Different methods of information processing are described. These include controlled processing, in which task performance is effortful and attention demanding. Tasks also become effortful if they have to be performed under stressful conditions (noise, loss of sleep, prolonged work etc). The cardiovascular measures that are sensitive to mental effort and to different forms of information processing are described, together with methodological problems that have to be taken into account. The usefulness of this approach in applied field settings is also outlined.

INTRODUCTION

Laboratory tasks (eg mental arithmetic, the Stroop test) are often used in cardiovascular psychophysiology. One reason for using them is to study the subject's reactivity to stressors. Strong reactivity (eg a large increase in blood pressure, heart rate or other cardiovascular parameter) is believed to be a risk factor for essential hypertension or coronary heart disease. In this type of research the nature of the task used is largely irrelevant. It is assumed that a subject's reactivity will become apparent in almost all cognitive tasks if these are made difficult enough. The only restriction is that the tasks should not be sensitive to differences between subjects in intelligence (see also Chapter 1). However, some cardiovascular psychophysiologists argue that the nature of the task is very important, and that the observed cardiovascular reactivity partly reflects the way in which the subject copes with the demands of this particular task. In this contribution we shall first discuss the processes underlying the performance of mental tasks and relate this analysis to the concept of coping. In the second part we shall discuss the analysis of cardiovascular responses and some methodological problems which are frequently encountered. Finally we shall discuss possible applications in field situations.

MENTAL TASKS

A mental task may be defined as a goal-directed activity pattern which is subject to certain environmental and structural constraints. The goals may be very different: to solve an arithmetic problem; to vocalise the colour of ink in which incompatible colour names are printed (eg to say 'green' when presented with the word 'red' printed in green ink); to search for the presence of a letter in a visual display containing letters or letters and numbers; to search for the presence of a letter in memory; to compare the meaning of a sentence to the meaning of a picture; to classify complex visual stimuli; to keep a mental account of how often certain environmental events occurred; to track a visual stimulus etc. These tasks may either be externally or self paced. Here we shall only discuss externally paced mental tasks. In order to understand the performance of the subjects it is necessary to have a theory of how subjects proceed in these tasks. It is assumed that performance in all these tasks depends upon three factors:

(a) The internal representation (IR) of stimuli. Internal representations may be present either in analogue form (eg, visual images) and/or in a more symbolic form. These internal representations (sometimes also called 'nodes') become complexly and increasingly interassociated and interrelated through learning. These nodes are normally passive and inactive and are called long-term store (LTS). The set of currently activated nodes is called short-term store (STS).

(b) Elementary information processes. The nodes can be activated by automatic or controlled operations. Automatic processing is a fast, parallel, fairly effortless process which is not limited by short-term store capacity. It is not under direct subject control and is responsible for the performance of well-developed skilled behaviours. Automatic processing typically develops when subjects process stimuli consistently over many trials. Controlled processing is characterised as a slow, generally serial, effortful, capacity-limited, subject-regulated processing mode which must be used to deal with novel or inconsistent information (Schneider et al 1984). Examples of automatic processing are semantic memory search and automatic encoding of stimuli. Examples of controlled processing are short-term memory search, rehearsal, decision processes, comparing, etc.

(c) The strategy of information processing. Subjects may differ in the sequence and nature of the elementary information processes they apply; they may also differ in the choice of the internal representation they use (ie, either analogue or symbolic). Therefore one way to characterise a strategy is by its component processes and the organisation of these in a coherent whole. Second, strategies are considered modifiable and flexible. In general, strategy refers to a set of internal cognitive procedures, a set that can be modified.

One reason for the observed physiological difference between subjects during the performance of cognitive tasks may be the strategy the subject uses in order to cope with the demands of the task. In other words, the cardiovascular state observed during a cognitive task may reflect the amount of time over which the subject is required to carry out attention-demanding information processing (Mulder 1980; Mulder and Mulder 1981). The time the subject needs for these processes is the time required (T_R); the time between successive signals is the time available (T_A). The subject can cope with the task if $T_A > T_R$, but the task becomes too difficult to cope with if $T_R \simeq T_A$ or $T_R > T_A$.

The availability of processing resources (either automatic or controlled) is believed to be dependent on the physiological state of the subject (eg Hockey and Hamilton 1983). For example, noise, anxiety and amphetamine may improve speed of performance, but also reduce accuracy. These 'stressors' are especially harmful for tasks largely depending on short-term store operations. Stressors such as noise, anxiety, amphetamine, nicotine, incentives, time of day, heat, loss of sleep, prolonged work and working under the critical presence of observers change the internal state of the subject, making some resources more and other less available. Without some executive control, these changes in availability would be readily observed in performance patterns. However, the subject may try to overcome the discrepancy between the current state (produced by the stressors) and the required state. The observed physiological activity during cognitive tasks may therefore also reflect another aspect of effort: the amount of executive resources the subject has to use to change the prevailing resource state in the direction of that which is optimal for the performance of a particular task (Hockey 1984; Sanders 1983). In this respect it is possible to speak of the <u>efficiency</u> of human information processing. Efficiency is decreased if the same performance level can only be obtained by investing more effort in the task; efficiency is increased if the same level of performance can be achieved with less effort. Effort is the central concept in coping with mental tasks. It has two different meanings: a task may become effortful if it requires attention demanding control operations and/or if the current resource state of the subject deviates from the required resource state to a marked degree.

PHYSIOLOGICAL INDICES

During and after the performance of mental tasks the activity in different physiological systems can be studied. The major challenge to the investigators of these measures is that they must establish resource specificity (ie, the sensitivity of the measures to the use of specific automatic or controlled processing operations). In case of specificity, a measure is <u>diagnostic</u> (eg, the physiological index indicates a load on perceptual resources). However, it is also possible that the measure reflects the aggregate demands imposed on all resources, in which case its diagnosticity is sacrificed for greater total <u>sensitivity</u>. This notion is conceptualised in Fig 1.

Activity patterns identified through assessments of regional cerebral metabolism and blood flow are believed to indicate the required cortical areas (resources), while components in cortical evoked potentials are believed to reflect the timing of perceptual, cognitive and motor events. In contrast, indices such as heart rate, blood pressure variability and pupil dilation are believed to be very sensitive to the controlled processing demands of cognitive tasks, but <u>undiagnostic</u> with regard to the specific locus of demand. Although a variety of cardiovascular parameters can be monitored in psychophysiological investigations, discussion here is mainly confined to phasic and tonic changes in heart rate, blood pressure and baroreflex sensitivity.

It is important to make a distinction between indices reflecting parasympathetic and sympathetic influences on the heart. It is argued that indices of phasic cardiac activity, in particular the amplitude of higher frequency components in the cardiac interval

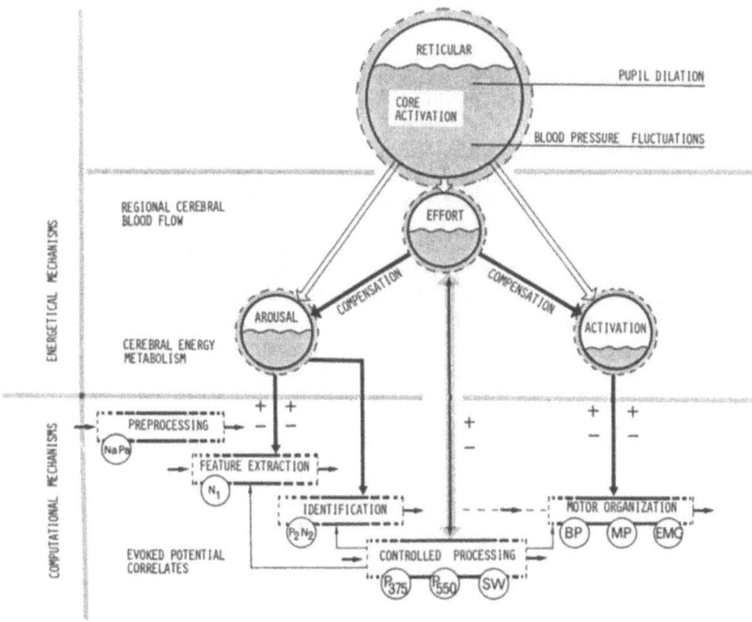

Fig 1. The connection between the structural and energetical components of the information processing system. The energetical resources are believed to have an hierarchical structure. The higher stages are considered to be less resource-specific. Arousal refers to a group of energetical resources related to encoding, activation refers to a group of energetical resources related to movement preparation and selection, while effort refers to energetical resources related to controlled processing.

spectrum and the amount of coupling between respiration and heart period, can serve as non-invasive indices of parasympathetic influence on the heart. The various parameters used in the assessment of sympathetic influences on the heart are discussed in detail elsewhere (Bunnell 1985; Lamberts et al 1985).

Computing Equidistant Time Series

The cardiac interval signal is a discontinuous signal. Actually it is a number of discrete pulses coincident with the R-waves in the ECG and these pulses are separated by unequal intervals. The computation of phasic cardiac activity measures and of certain tonic measures requires an equidistant interval series. Four alternative methods which aim to obtain an equidistant time series are currently found in the literature (see Mulder 1980 for a review): interpolation, low-pass filtering, the interval tachogram and the cardiotachogram. We recommend the use of low-pass filtering (see also Chapter 14). This method is based upon a technique of alias-free sampling of neuronal spike trains. The resulting time series reflects an equidistant time series of instantaneous heart rate, and can be expressed in terms of the modulation index (MI), ie the ratio between variations in heart rate and the mean value (see Mulder L 1985).

Phasic Cardiac Activity

When experiments are conducted with paced mental tasks, the subject is required to process information during the interval between two successive imperative signals. With the use of averaging procedures it is possible to establish whether there is systematic phasic cardiac activity during this interval. The acquisition of averaged response patterns is detailed in Chapter 13.

Figure 2 shows the results of such a procedure applied to a sentence comprehension task. The subject is required to compare a statement about a situation (eg, 'It is not true that the dots are red') with a picture (eg, black dots).

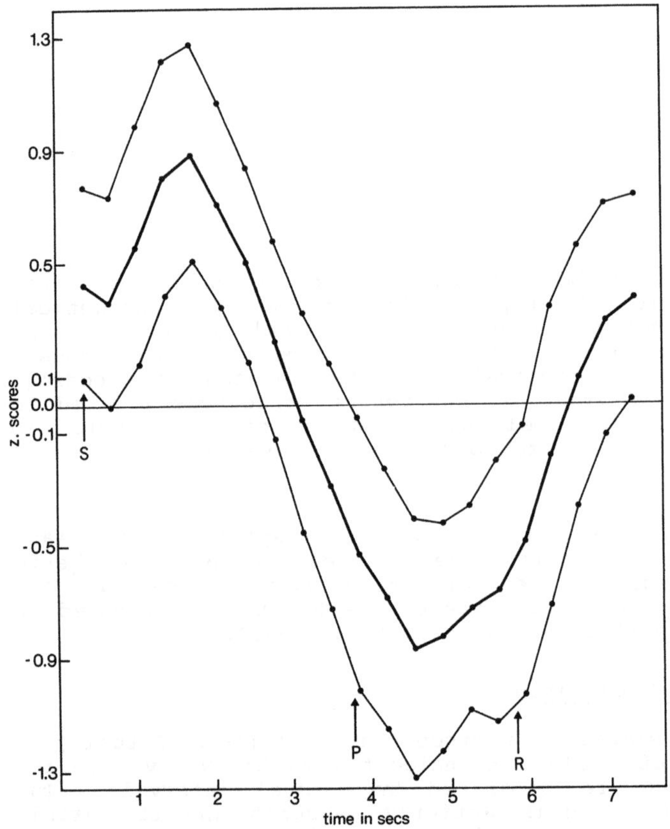

Fig 2. Evoked heart rate response expressed in two scores obtained during a difficult sentence comprehension task. Heavy line: mean value, thin lines: confidence interval (95%). S indicates the moment the sentence was presented; P indicates the mean reaction time obtained in this condition.

The brief heart rate deceleration immediately after the

presentation of the sentence (S) reaches its nadir at about 500 msec. Subsequently, heart rate starts to accelerate and a maximum is reached about about 1,500 msec after presentation of the sentence. From that point heart rate decelerates, and acceleration starts shortly after the presentation of the picture (P). These phasic changes in heart rate occur within a period of 7,400 msec (the time between two successive sentences) and probably reflect the moment-to-moment use of cognitive resources. The sentence has to be encoded and an internal representation has to be formed (deceleration, acceleration); then the subject starts to prepare for the picture (deceleration), forms an internal representation of the picture and compares it with the internal representation of the sentence (acceleration). These relatively fast changes are believed to be controlled by the vagus, since Obrist and colleagues (1974) have shown that these phasic heart rate changes are relatively unaffected by sympathetic blocking agents. A more detailed picture can be obtained by determining the point in the cardiac cycle at which signals are presented. The duration of the cycle is maximally prolonged if the imperative signal is presented soon after the R-wave (see Chapter 13).

The importance of these patterns to cardiovascular pathology is reflected in the observation that vagal activation may be a necessary condition for the induction of atrial fibrillation (Yelich et al 1978). On the other hand, joint parasympathetic-sympathetic activation has been related to sudden death. Glass (1977) suggested that the Type A individual experiences periods of intensive effortful coping alternating with periods of withdrawal and hopelessness. These states are accompanied by periods of high catecholamine excitation and inhibition respectively. Such periods of alternating vagal and sympathetic activation can also be observed in the cardiac evoked response during the performance of the cognitive task. In addition, the vagus may also exercise a protective function. Establishing reliable individual differences in the vagal control of the heart would therefore appear to be a major step in identifying individuals at risk. The use of cognitive tasks in which vagal modulation of cardiac activity is reliably observed can be highly recommended in epidemiological research.

Tonic Cardiac Activity

If a task lasts several minutes, cardiac activity can be described in tonic terms. The most straightforward way is to describe the signals in the time domain (eg, mean cardiac interval time, mean blood pressure, mean pre-ejection period, or the variance of these measures). Spectral analysis, on the other hand, provides a more detailed picture of the fluctuations in these signals within a certain time period by describing the frequency content. Figure 3 illustrates spectra of systolic blood pressure and the cardiac interval signal. If the time period is sufficiently long (about ten minutes) and the signals are reasonably stationary, it is possible to observe three major spectral components in both signals: one at about 0.04Hz (the low-frequency component), one at about 0.10Hz (the mid-frequency component) and one corresponding with the dominant respiratory frequency of the subject.

There are two theories concerning the origin of the lower spectral components. Hyndman et al (1971) have argued that the 0.10Hz component originates from a system controlling arterial pressure. The control system structure is believed to be of a negative

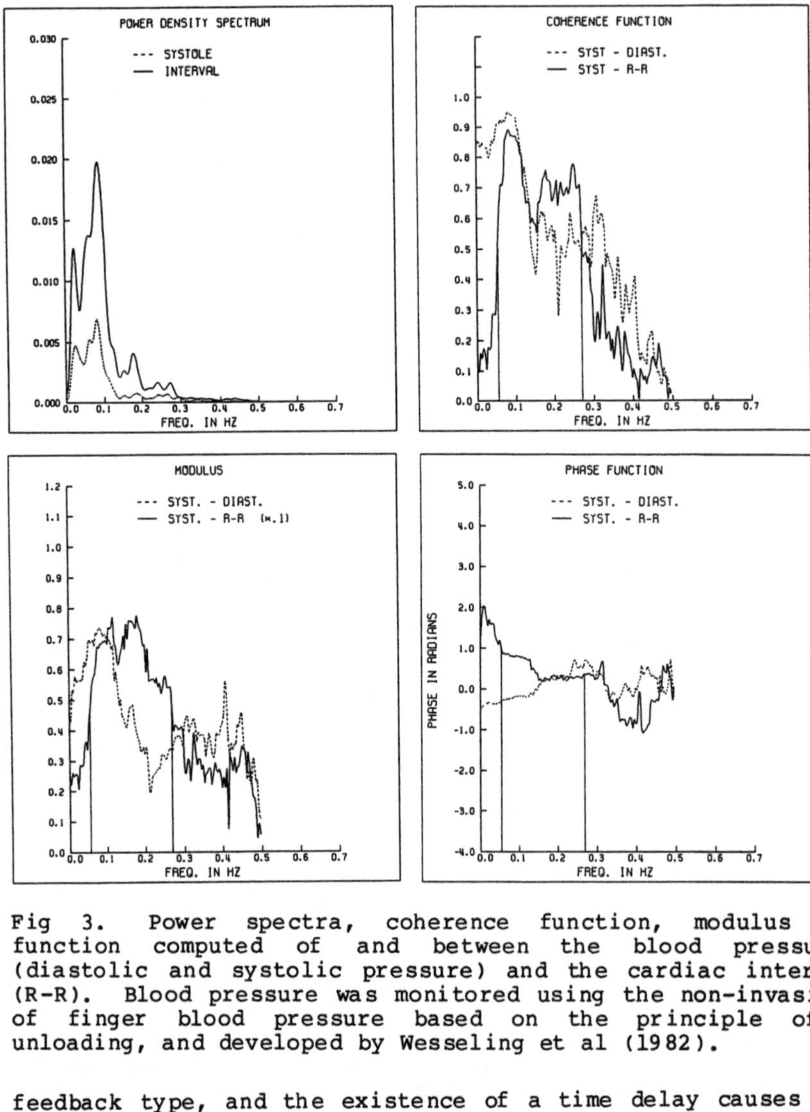

Fig 3. Power spectra, coherence function, modulus and phase function computed of and between the blood pressure signal (diastolic and systolic pressure) and the cardiac interval signal (R-R). Blood pressure was monitored using the non-invasive measure of finger blood pressure based on the principle of vascular unloading, and developed by Wesseling et al (1982).

feedback type, and the existence of a time delay causes the system to oscillate. This occurs if there is a delay of a half-cycle and the gain is greater than one (Mulder 1980; Kitney and Rompelman 1980).

Wesseling and Settels (1985) on the other hand argue that the 0.10Hz component represents the resonance frequency of the baroreflex. The amplitude of the resonance frequency increases with increasing total baroreflex loop gain. Akselrod et al (1981) demonstrated that parasympathetic blockade abolishes the mid- and high-frequency components. On the other hand, a blockade of the renin-angiotensin system increases the low frequency component. If we assume that this component originates from fluctuations in peripheral vasomotor tone, then the renin-angiotensin system appears to damp the amplitude of these fluctuations. It has been suggested that the spectrum of both signals can usefully be divided into three frequency bands:

(a) the low frequency band (from 0.02 to 0.06 Hz);
(b) the mid frequency band (from 0.07 to 0.14 Hz);
(c) the high frequency band (from 0.15 to 0.50 Hz).

In a number of studies, we have found the amplitude of the mid frequency component to be systematically decreased if the task required more controlled processing (Mulder 1980; Mulder and Mulder 1981). The results of a recent experiment are outlined in Fig 4 (Aasman et al, in preparation). In this experiment, subjects were confronted with two levels of a memory search task. In the 'simple' condition subjects had to determine whether or not a letter was a member of a previously presented memory set. In the 'double' condition the subject in addition had to keep a mental account of how often each letter of the memory set was presented. As the interval between stimulus presentations (T_A) becomes increasingly filled with mental operations (T_R), the amplitude of the mid-frequency component decreases systematically. When a running count of five is required, the task becomes too difficult, error rate increases, and vagal control of the heart also increases.

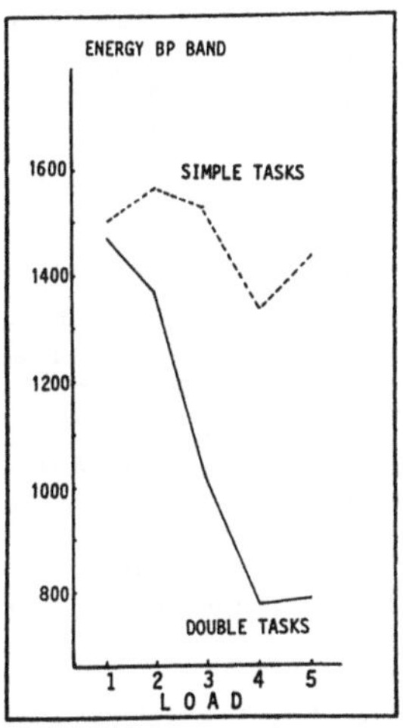

Fig 4. Power of the 0.10Hz component of the cardiac interval signal in a short-term memory task.

Using spectral methods, it is also possible to compute the coherence between cardiac interval and blood pressure signals. Of special importance is the weighted coherence function. This indicates the proportion of the total variance of one signal that is shared with another signal within a certain frequency band. From Fig 3 it is clear that there is a high coherence between blood pressure and heart rate between 0.05 and 0.30 Hz. In this frequency range the heart rate component of the baroreflex is active. The amount of coupling between the signals can be further explored by computing the modulus and phase of the transfer function (also shown in Fig 3). The phase expresses the time difference between the two signals as a function of frequency. This time difference reflects the total time from baroreceptor activation to vagal control of the cardiac pacemaker. The modulus (or gain) between systolic pressure an the interbeat interval gives an indication of the transfer from mmHg in systolic pressure to msec in interval time (in this case about 10msec/mmHg). The modulus of the transfer function can be considered as an index of the sensitivity of the heart rate component of the baroreflex. Baroreflex sensitivity has been shown to decrease as human subjects are taken from sleep through progressively higher levels of mental activity (Conway 1983). The observed decrease in amplitude of the 0.10Hz oscillations in the cardiac interval signal as a function of the controlled processing demands of the task therefore probably reflects a graded inhibition of the (heart rate component of the) baroreflex. There may be large individual differences in the central modulation of the baroreflex responsible for individual differences in cardiovascular state during the performance of mental tasks. Recently Sleight (1983) showed that individual differences in baroreflex sensitivity predict pressor responses to a variety of stressors, including a reaction time task and mental arithmetic.

Computing Baroreflex Sensitivity

It is clear that the sensitivity of the baroreflex may be a very important psychophysiological index. Until now two techniques have mainly been used to study arterial baroreflex sensitivity during mental task performance. The first is to study the reflexogenic changes in heart rate during brief periods of drug-induced (mostly phenylephrine) blood pressure elevation (Conway 1983). However, one of the problems with the phenylephrine technique is that it does not provide adequate information about the control of blood pressure since part of the pressure rise is induced by the drug itself.

The second technique is to change the transmural pressure over the carotid sinus (via neck suction) and to analyse the reflexogenic changes in heart rate and blood pressure. Recently, Forsman and Lindblad (1983) used this technique to evaluate baroreflex sensitivity during the Stroop test (a colour-word conflict test). The results showed that both heart rate and blood pressure increased during the task, but that there was no change in baroreflex sensitivity. It is possible that the difference in cognitive demands between mental arithmetic and colour-word conflict tasks may be responsible for this discrepancy (Steptoe 1985).

Steptoe (1981, 1985) has argued that it is possible that inhibition of short-term haemodynamic control through baroreceptor reflexes facilitates exaggerated pressor responses to behavioural stressors in vulnerable individuals. If this activation pattern occurs repeatedly, a state of fixed hypertension may eventually occur. The baroreceptor reflex may then be reset to higher values. As we have seen in the preceding section, the amplitude of the 0.10Hz component decreases as the task requires more controlled processing. Such a decrease can be explained by a decreased gain of the baroreflex. Hypertensives appear to have a more permanently decreased sinusarrhythmia, probably reflecting a more permanent change of baroreflex sensitivity. Clearly it would be most desirable to develop a non-invasive index of baroreflex sensitivity. Such an index should be sensitive to both short-term (within a minute) and long-term (minutes, hours) changes in baroreflex control.

Recently, De Boer et al (1983) and Karemaker (1985) have suggested a set of difference equations describing beat-to-beat values of the n-th heart period (I_n), the pulse pressure (P_n), diastolic pressure (D_n), and peripheral resistance (R_n), all observed during I_n. Although other factors (such as the Windkessel effect and Starling's Law) have been accounted for in supplementary formulae, the equation given below describes the fast, vagally mediated, baroreflex effect of the n-th systolic pressure on the duration of the n-th interval. If resting heart period is shorter than about 800 msec, one may find $\alpha=0$ and α_1 in operation, at longer heart periods α_0 is in operation and $\alpha_1=0$.[1]

$$I_n = \alpha_0 S_n + \alpha_1 S_{n-1} + C_1$$

We have used this mathematical model to compute baroreflex sensitivity, α, over periods of five minutes, and for every 20 seconds. Such procedures can be very useful in psychophysiology (see Mulder L 1985; Veldman et al 1985). However, the reader must realise that until now the interaction between the respiratory system and the cardiovascular system has not yet been incorporated into the mathematical models.

METHODOLOGICAL ISSUES

Having computed phasic and tonic measures of cardiovascular activity the investigator has to face a number of methodological problems. Some of the problems which are most often encountered are listed below.

1. Differences between subjects in performance and the effects of practice.
2. Separating the effects of habituation from task induced effects.
3. Controlling for the indirect effects of related physiological (motor) activity.
4. Dealing with the multivariate nature of cardiovascular responses.

Many of these issues are discussed elsewhere in this volume, but brief mention will be made of the methods we have adopted to resolve them.

Differences Between Subjects in Performance and the Effects of Practice

Even in simple tasks there are differences between subjects in their performance, eg, number of correct responses, mean reaction time. These individual differences may reflect the different strategies with which subjects approach the task, producing different patterns of cardiovascular activity. One of the first steps the experimenter should take is to identify systematic individual differences in performance. Such an approach uses the similarities between the subjects' patterns of performance in the experiment as the basis for separating them into subgroups (mini-types; see Kareev, 1982). Having identified mini-types, these subgroups can be introduced into the analysis of variance. Strong mini-types by task variable(s) interactions, or a mini-type main effect indicate the existence of systematically different responses.

A related problem concerns the effects of practice. If the task requires practice, the researcher first has to establish if practice was enough. It is most unpleasant to discover that subjects still improve in performance across experimental conditions. One possibility is to use varied mapping conditions (Schneider et al 1984). If the stimuli and responses are variably mapped across trials (that is, the responses should change across trials) no automatic processing should develop and task performance remains effortful.

Separating the Effects of Habituation from Task Induced Effects

The second problem with which the experimenter is confronted is that a task presented in the beginning of an experiment induces a stronger level of activation than the same task presented later in the session. In order to cope with this problem, the order of task presentation can be varied, and experimental tasks can be repeated several times before the actual measurement. We have used the following procedure:

Order 1. Rest - Ref 1 - Easy - Ref 2 - Difficult - Ref 3
Order 2. Rest - Ref 1 - Difficult - Ref 2 - Easy - Ref 3

We start with a rest period, followed by the first reference (Ref) task. Reference tasks are very simple versions of the main task. Rest is a very undefined state, in which one does not control the subject's internal activity. The reference tasks require some attention of the subject, and the amount of motor activity is the same as the main tasks. The two main tasks (Difficult and Easy) differ in the amount of controlled processing they require, and the difference between these two versions may be affected by habituation. If order is introduced as a factor in the analysis of variance it is possible to explore its presence as a main effect or its interaction with task difficulty. This design also makes it possible to observe after-effects of tasks. If a difficult task affects the following reference task, a trend by order interaction should be expected.

Controlling for the Indirect Effects of Related Physiological Activity

The third problem is very difficult to solve. If, for example, one compares a rest condition with a Stroop test, the observed difference in cardiovascular activation may result from task-induced effects, but also from a change in respiratory pattern (the subjects are required to mention the colour of a word). Similarly, using reaction time tasks, the rest and task conditions differ in the amount of motor responses. It is for that reason that we decided to keep the amount of motor activity constant: same rate of signal presentation and rate of responding across reference and main tasks. Direct comparisons can then be made, and these are not compromised by indirect effects.

The Multivariate Nature of Cardiovascular Responses

An experiment in cardiovascular psychophysiology usually results in a large number of physiological responses. In principle it is possible to analyse the results by using a series of univariate tests for each variable. For each of these variables it is worthwhile to explore the sensitivity of the experimental design. A quantitative index of the sensitivity of an experiment is its power. Power refers to the possibility of rejecting the null hypothesis when an alternative hypothesis is true. The power of a test is directly related to α, the level of significance testing: an increase in the size of α will result in an increase in power. However, there are also other ways on increasing the power of an experiment:

Decrease in the uncontrolled variance: Power is increased if we are able to eliminate uncontrolled sources of variance. An increase in the size of the error variance reduces the size of the F ratio and lessens our chances of rejecting the null hypothesis. There are three major sources of error variance: random variations in the actual treatments, the presence of unanalysed control factors and individual differences. The first source can be dealt with by a strict standardisation of the measurement procedure. The second source may arise for example from habituation or carry-over effects, and may be controlled by introducing order into the design. The effects of individual differences have already been discussed. Finally, most researchers use a repeated-factors design. In general the error term associated with repeated factors (within-factors) are considerably smaller than those with non-repeated factors (between-factors). Unfortunately, this method has rather restrictive assumptions of homogeneous variances and covariances (see below).

Increase the treatment effects: Power is also increased as the size of the treatment effects is increased. This may be accomplished by choosing widely spaced levels of the independent variable or by introducing conditions which are thought to maximise the appearance of the phenomenon (eg, to introduce noise during a mental arithmetic task; see also Chapter 4).

Increase the number of observations per treatment condition: Power is also increased as the number of observations per treatment condition is increased.

In principle it is possible to analyse the results of a cardiovascular psychophysiology experiment by using a series of univariate tests for each dependent variable. However, since each of the univariate tests is designed to produce a significant result ($\alpha \times 100\%$ of the time when the null hypothesis is correct), the probability of having at least one of the tests produce a significant result by chance increases rapidly as the number of tests increases. It is thus highly desirable to have means of explicitly controlling error rate. Multivariate analysis of variance provides this control. In fact the effects of all variables are tested simultaneously as a vector. Another important possibility of multivariate analysis is the use of step-down analysis. Using this procedure, the dependent variables are tested in a special order, and the significance test of the first variable is identical to the univariate test. The significance of the next variable is tested after the effect of the first variable is partialled out, etc. Finally, multivariate analysis of variance does not have the highly restrictive assumptions about the nature of the inter-relations among subjects' responses to the various levels of the within-factor.

The multivariate approach sketched above is static. It is based upon mean values and their intercorrelations. However, the moment-to-moment adaptation of the cardiovascular system to the demands of the task are also important and may be most appropriately described within models of circulatory dynamics (see

Wesseling and Settels 1985). Such procedures may provide more insight into the complex interactions within the system than most of the current type of experiments in cardiovascular psychophysiology do.

FIELD APPLICATIONS

Is it possible to use this approach in more applied situations? We think that it is, and recently carried out a series of studies of city bus drivers. City bus drivers experience an unusual burden of occupational stress as evidenced by their average absenteeism (17-25%, or twice the Dutch yearly industrial mean). An epidemiological study of absenteeism and turnover among drivers operating within the city of Groningen over a 15-year period (1964-1978) revealed that only one out of ten drivers leaving the company during this period had reached the official retirement age, 60 years. Six out of ten drivers retired early for reasons of medical disability, at an average age of 47 years (Mulders et al 1982).

In order to obtain a more precise insight into this unhappy state of affairs we decided to divide a selected group of drivers into three subgroups: those with a high, medium and low sickness leave. After a short training period we measured their mental task performance and cardiovascular activity on two work-days and two free days. On the work-days the measurements were taken before driving, after three hours and after six hours of driving. The tasks we used ranged from relatively simple to very complex. In the simple condition the subjects were required to memorise either two or four items; after having done this they were presented with a series of 40 items and had to decide each four seconds (by pressing a 'yes' or a 'no' button) whether or not a presented item was a member of the previously memorised set. Each day a different memory set was used and consequently the task could not be automatised. There were two relatively difficult conditions. In one of them the memory set was four and the drivers had to count the number of times each memory set item was presented. This task was judged to be very attention-demanding and effortful. In the other condition the drivers had to combine the reaction time task with counting. This condition was judged to be extremely effortful. The whole test lasted 20 minutes. During all these conditions the ECG was recorded, and the amplitude of the 0.10Hz computed from the spectra.

By using this design it was possible to establish whether the preceding driving period affected the availability of processing resources (increased search and decision times, decreased efficiency to time share) or that performance remained virtually unaffected but at the cost of more effort (a more suppressed 0.10Hz component). It appeared that in all drivers mental efficiency decreased and effort (indexed by the difference between rest and task) increased. But sickness level appeared to be a particularly important determinant. Using the most difficult task it was possible to classify the drivers very accurately (90%). Drivers with a medium sickness level appeared to increase their reaction times and number of errors substantially. Apparently they were unable to compensate since their 0.10Hz component increased considerably, indicating a higher parasympathetic control of the heart. In contrast, drivers with a low sickness level remained very efficient, but also showed a suppressed 0.10Hz component.

These effects are summarised in Fig 5. Using this design it will become possible to assess the amount of workload in different occupations and to identify individuals at risk.

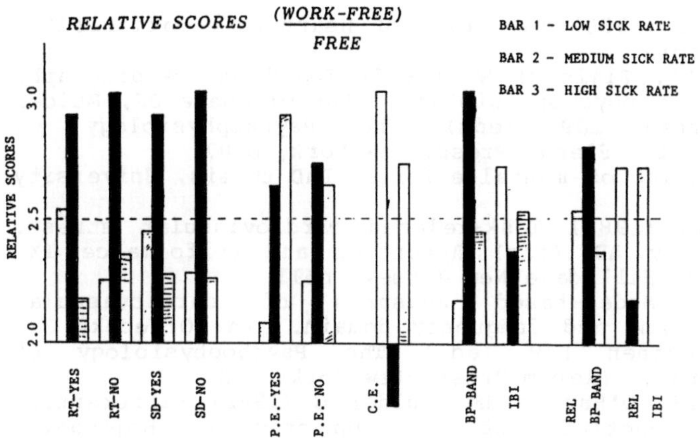

Fig 5. Relative scores (Work-free/Free day performance) of city bus drivers with different sickness rates.

REFERENCES

Akselrod S, Gordno D, Ubel FA, Shannon DC, Cohen RJ (1981) Power spectrum analysis of heart rate fluctuation: a qualitative probe of beat-to-beat cardiovascular control. Science 213: 213-220

Bunnell DE (1985) Noninvasive measurement of sympathetic influences on the heart. In: Orlebeke JF, Mulder G, van Doornen LJP (eds) The Psychophysiology of Cardiovascular Control. Plenum Press, New York, p221

Conway J, Boon N, Vann Jones J, Sleight P (1983) Involvement of the baroreceptor reflexes in the changes in blood pressure with sleep and mental arousal. Hypertension 5: 746-748

De Boer RW, Karemaker JM, Strackee J (1983) Beat-to-beat variability of heart interval and blood pressure. Automedica 4: 217-272

Forsman L, Lindblad LE (1983) Effects of mental stress on baroreceptor-mediated changes in blood pressure and heart rate and on plasma catecholamines and subjective responses in healthy men and women. Psychosom Med 45: 435-445

Glass DC (1977) Behavior Patterns, Stress and Coronary Disease. Lawrence Erlbaum Ass, Hillsdale New Jersey

Hockey GRJ, Hamilton P (1983) The cognitive patterning of stress states. In: Hockey GRJ (ed) Stress and Fatigue in Human Performance. Wiley, New York, p331

Hockey GRJ (1984) Varieties of attentional state: the effects of the environment. In: Parasuraman R, Davies DR (eds) Varieties of Attention. Academic Press, London, p449

Hyndman BW, Kitney RI, Sayers BMcA (1971) Spontaneous oscillations in physiological control systems. Nature 233: 339-341

Kareev Y (1982) Minitypologies from within-subjects designs: uncovering systematic individual differences in experiments. J Verb Learng Verb Behav 21: 363-382
Karemaker JM (1985) Short-term regulation of blood pressure and the baroreflex. In: Orlebeke JF, Mulder G, van Doornen LJP (eds) The Psychophysiology of Cardiovascular Control. Plenum Press, New York, p55
Kitney RI, Rompelman O (1980) The Study of Heart Rate Variability. Clarendon Press, Oxford
Lamberts WM, Visser KR, Zijlstra WG (1985) Impedance cardiography and its usefulness in psychophysiology. In: Orlebeke JF, Mulder G and van Doornen LJP (eds) The Psychophysiology of Cardiovascular Control. Plenum Press, New York, p207
Mulder G (1980) The heart of mental effort. PhD thesis, University of Groningen
Mulder G, Mulder LJM (1981) Task-related cardiovascular stress. In: Long J, Baddeley AD (eds) Attention and Performance IX. Lawrence Erlbaum Ass, Hillsdale New Jersey, p591
Mulder LJM (1985) Model-based measures of cardiovascular variability in the time and frequency domain. In: Orlebeke JF, Mulder G, van Doornen LJP (eds) The Psychophysiology of Cardiovascular Control. Plenum Press, New York, p333
Mulders HPG, Meyman TF, O'Hanlon JF, Mulder G (1982) Differential psychophysiological reactivity of city bus drivers. Ergonomics 25: 1003-1011
Obrist PA, Howard JL, Lawler DE, Galosy RA, Meyers KA, Gaebelin CJ (1974) The cardiac-somatic interaction. In: Obrist PA, Black AH, Brener J, Dicara LV (eds) Cardiovascular Psychophysiology. Aldine, Chicago, p136
Sanders AF (1983) Towards a model of stress and human performance. Acta Psychologica 53: 61-69
Schneider W, Dumais S, Shiffrin RM (1984) Automatic and control processing and attention. In: Parasuraman R, Davies DR (eds) Varieties of Attention. Academic Press, London, p1
Sleight P (1983) The influence of arterial baroreceptors in man on the variability of blood pressure and plasma catecholamines in man. Chest 83: 320-322
Steptoe A (1981) Psychological Factors in Cardiovascular Disorders. Academic Press, London
Steptoe A (1985) Behavioral influences on the mechanisms of hemodynamic regulation. In: Orlebeke JF, Mulder G, van Doornen LJP (eds) The Psychophysiology of Cardiovascular Control. Plenum Press, New York, p169
Veldman JBP, Mulder LJM, Mulder G, van der Heide D (1985) Short-term coherence between blood pressure and heart rate during mental loading: an exploration in the time and frequency domain. In: Orlebeke JF, Mulder G, van Doornen LJP (eds) The Psychophysiology of Cardiovascular Control. Plenum Press, New York, p391
Wesseling KH, De Wit B, Settels JJ, Klawer WH, Arntzenius AC (1982) On the indirect registration of finger blood pressure after Penaz. Funkt Biol Med 1: 245-250
Wesseling KH, Settels JJ (1985) Baromodulation explains short-term blood pressure variability. In: Orlebeke JF, Mulder G, van Doornen LJP (eds) The Psychophysiology of Cardiovascular Control. Plenum Press, New York, p69
Yelich MR, Euler DE, Wehrmacher WH, Sinha SN, Randall WC (1978) Parasympathetic influence on atrial vulnerability in the puppy. Am J Physiol 235: H683-H689

CHAPTER 4

The Quantification of Experimental Load: Methodological and Empirical Issues

Werner Kuhmann, Harald Lachnit and Dieter Vaitl

Fachbereich Psychologie, Justus-Liebig-Universitaet Giessen, Otto-Behaghel-Strasse 10, D-6300 GIESSEN, Federal Republic of Germany

SUMMARY

One problem in psychological stress testing is that the load characteristics of an applied set of experimental conditions are difficult to evaluate. A second problem is that subjects can modify the characteristics of the test situation through their behaviour. Two solutions of these problems are suggested. Firstly, the manipulation of test conditions with respect to a single, common dimension (such as test difficulty); secondly, the individual standardisation of test situations with respect to a common dimensional variable. These approaches are illustrated in two experiments. The results suggest that both procedures are effective. It is possible to establish load conditions with specific quantitative characteristics (task difficulty) which are reflected by behavioural dependent variables. The way in which subjects cope with experimental tasks can then be evaluated through monitoring physiological and emotional parameters, which in this model are regarded as moderating variables.

INTRODUCTION

The main issue underlying this contribution is that in cardiovascular stress research, it is not sufficient simply to assess the cardiovascular responses of subjects. It should also be possible to relate these responses to the difficulty (or power) of the applied test, and to the behavioural performance of the subject. Although this chapter concerns mental stress testing in cardiovascular research, the rules governing the analysis of test conditions are general.

Our guiding principle is that it is essential to test subjects over a range of predetermined levels of mental load. These different loads should represent levels of a dimensional variable, such as difficulty of a single task, so that different test conditions can be arranged in an unequivocal order. These levels should be reflected in scores on behavioural dependent variables.

The importance of this factor is apparent in contemporary cardiovascular stress testing. When mental arithmetic, sentence completion, noise, vigilance tasks, the cold pressor and other tests are administered, they certainly represent different levels of load. But the whole character of task varies from one situation to another with respect to intensity and quality. The order of the test situations (in terms of mental load) is normally deduced from the magnitude of cardiovascular responses. But then the problem arises that in a second test series, or a different group of subjects, the ranking of tasks may change because of variations in physiological response amplitude. For example, what conclusion can be drawn if a subject produces a heart rate of 140 bpm on the first presentation of a test, and 80 bpm on the second? Can it be concluded that only the first presentation is valid, or should we conclude that the subject has learnt something about the test and the conditions under which it is presented? What conclusions can be drawn concerning the cardiovascular risk of subjects in these situations? In our opinion, these questions can only be answered if the subject's behavioural performance is assessed, and if the characteristics of the test (both physical and psychological) are the same on both presentations.

A simple general model is proposed, such that in any test situation, some specific parameters of the situation act upon the person (eg, a specific test is presented). The person interacts with the situation, and tries to solve the task adequately. Such performance requires the activation of appropriate personal capacities which might be physical, mental and social. This interaction between person and situation will result in a certain level of performance, and this can be used to differentiate between people confronted with the same test under similar conditions. In addition, people with the same behavioural performance can differ with respect to their level of activation, and this may be reflected in cardiovascular responses.

In this model, the specific parameters of the situation can be regarded as independent variables (IV), the performance on the task is the dependent variable (DV), and the activation parameters are moderating variables (MV). The dependent variable should be a specific response to the independent variable; for example, arithmetic performance in a mental arithmetic task or work output in an ergometer test, and not heart rate or blood pressure. The independent variables control the dependent variable. However, this effect is mediated by the attributes and behaviour of the subject. In cardiovascular stress testing, moderating variables such as heart rate, blood pressure and catecholamines are of special interest. But a valid interpretation of changes in moderating variables depends on the properties of the experimental task, as well as on the subject's behaviour and performance. To avoid circularity, information about the experimental load must be independent of the actual dependent and moderating variables. Furthermore, the test must have stable attributes, and not change markedly on repetition.

Unfortunately, it is virtually impossible to assess and control all the components of a test situation. Subjects are confronted with many different demands; the independent variables, social demands, environmental disturbances, novelty, incentives and disincentives, subjective importance of the test, etc. Some of these features change during test administration, while others are stable. Valid interpretation of results is facilitated by excluding as many unstable or unknown components from the test situation as possible. Furthermore, different test levels should be achieved by manipulating only a few parameters, rather than comparing completely different tasks. In formal terms, quantitative levels of the independent variables should be selected.

Two aspects of test situations appear particularly important in mental stress testing:

1. The relative difficulty of the task, expressed in quantitative terms.

2. The performance of a subject on a series of task items.

These aspects were investigated in the two experiments outlined here. The aim of both experiments was to show that variations of levels within the independent variables resulted in corresponding performance differences, and that the effects of between- subject variation could be minimised. We suggest that this kind of testing may be used to provide a reliable basis on which to interpret cardiovascular behaviour. However, in order to realise our primary aims, no cardiovascular data were recorded in Experiment 1, although heart rate was assessed as a moderating variable in Experiment 2.

EXPERIMENT 1

The main independent variable manipulated in this study was the difficulty of a concentration task; this was set at five levels. However, to test whether the effects of task difficulty are influenced by the way items or levels are presented, two further aspects were varied: order of presentation of the different levels of task difficulty (two orders), and knowledge of task performance (two kinds of feedback and a no-feedback condition).

The task items were taken from the 'Konzentrations-Verlaufs-Test' (Abels 1965). Each item consists of 36 two-digit numbers presented in a matrix of six rows and six columns. The task is to detect in which combinations the numbers 43 and 63 occur in the matrix. The items are simple, so only low intellectual capacities are required for task solution. In this experiment the task was slightly changed from the original version, so that the four combinations, 43 only, 63 only, 43 and 63, and neither 43 nor 63, have equal probabilities. Task difficulty was manipulated by varying presentation times. Five trials, each of 30 items, were administered, with presentation times of nine, six, five, four and three secs respectively. Within each trial, all 30 items had the same presentation time. Prior to these experimental trials, a practice trial with 12-sec presentation was administered.

The order of presentation was manipulated, so that for half the subjects the trial order varied at random (permutation condition 'PER'), while the remaining subjects were presented with trials of increasing task difficulty ('INC' condition). Subjects were informed about the order of presentation. Thus in the INC condition, subjects were able to anticipate the difficulty of the next trial and to prepare themselves in advance. The knowledge of performance manipulation involved three conditions: success feedback, failure feedback, and no feedback. In the feedback conditions, a tone was presented immediately following the behavioural response, or at the expiry of the presentation time. Forty-two male and forty-two female students were assigned to six experimental groups (combining the two orders of presentation and three feedback conditions).

The results of the study, expressed in terms of the percentage of correct responses, are illustrated in Fig 1. It can be seen that the levels of task difficulty were strongly associated with behavioural performance. High levels of task difficulty were

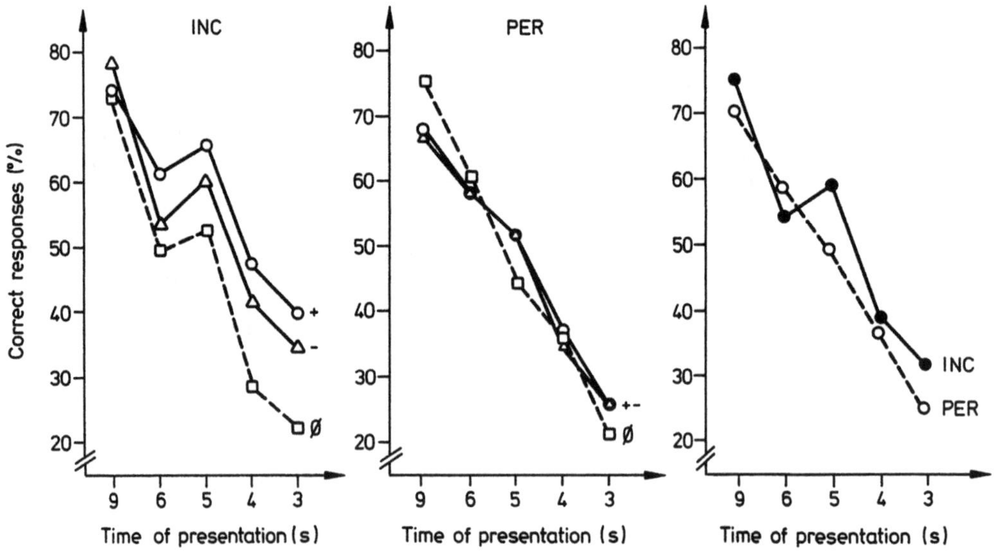

Fig 1. Percentage of correct resonses for the different types of feedback and the order of presentation, Experiment 1 (left and middle). The right part shows the means for order of presentation averaged over all three types of feedback.

+ = success feedback
- = failure feedback
0 = no feedback
INC = increasing order of task difficulty
PER = random order of task difficulty

associated with poor performance, and this effect was independent of the order of presentation (see right panel of Fig 1). At first sight, this result may seem trivial. But it should be noted that it confirms that the postulated levels of the independent variable represent different values on a dimensional parameter, and that these were defined independently of subjects' performance.

The order of presentation variable modulated performance, and this was reflected in a significant interaction between order of presentation and task difficulty. The INC condition evidently permitted subjects to prepare for the difficulty of forthcoming trials, and at medium levels of task difficulty (5-sec presentation), this led to different performance levels.

Self-report data suggest that the knowledge of performance affected the emotional state of participants. Subjects in the failure feedback conditions (PER-, INC-) rated the feedback tone as less pleasant than those in the success feedback condition (PER+, INC+). Despite this, there were no differences in performance scores. Nevertheless, the feedback manipulation controlled another moderating variable, namely the degree of engagement in the experiment. Subjects in both feedback conditions omitted fewer items than those without feedback. This may account for the observation that subjects without feedback performed less well than those in the feedback conditions, although only in the three INC groups (Fig 1, left panel).

EXPERIMENT 2

The second experiment differed from Experiment 1 in that the performance on the task was used as the independent variable (Walschburger et al 1980; Lachnit and Meinardus 1981). The levels of the independent variable were defined operationally in terms of different degrees of failure. The same tasks were used as in Experiment 1. Two levels were studied. These were a low percentage of failure condition (realised by using long presentation times), and a 50-60% failure level (realised by an algorithm that adjusted presentation times according to individual subjects' work rates). These conditions were intended to induce different amounts of failure, but similar amounts of failure within each level (Kuhmann 1979).

The experiment consisted of three situations: (1) a low percentage of failure, induced by long presentation times of 30 secs per item for four minutes; (2) 50-60% failure rate for seven minutes; (3) low percentage of failure again for four minutes. All subjects were given acoustic failure feedback to enhance their engagement in the task. Twenty-four male and 16 female students participated in the study.

The results indicate that individual standardisation of experimental loads was successful, as can be seen in Table 1. The success of this manipulation of the independent variable is endorsed by the low subject variance relative to total variance. The dependent variable (reaction time) might appear to be confounded with independent variables in this experiment, because the experimental algorithm connects failure with presentation times, and hence indirectly with reaction time. But when the percentage of failure is fixed by the algorithm, both reaction time and presentation time vary with the subject's behaviour.

If reaction times are reliable dependent variables, then between-subject variance should be substantial in relation to total variance. The variance accounted for by experimental situation should also be large if level of failure controls reaction times. In fact, the between-subject variance was about one third (31%) of the total variance in reaction times, while situations accounted for 55%. As the means in Table 1 indicate, reaction times were slower in Situation 1 than 3, even though these were equivalent with respect to experimental load. Subjects were working more effectively in Situation 3, presumably because training acted as a moderating variable.

Heart rate data are presented in Table 1. Analysis of raw scores indicated that different heart rates were observed in the three situations, all of which differed significantly from each other. Subjects showed lower heart rate in Situation 3 than in Situation 1; coupled with their faster reaction times, this suggests that participants were coping more effectively with experimental demands. The differences between subjects are reflected in the fact that 87% of the heart rate variance was accounted for by between-subject variance, although the subjects by situations interaction was also significant. It is possible to argue from

Table 1. Means and standard deviations for the variables percentage of failure, reaction time and heart rate. Estimation of variance components for percentage of failure are also provided.

	Percentage of failure		
	Situation 1	Situation 2	Situation 3
mean	16.44	57.86	16.23
sd	9.87	3.93	12.76

Estimation of variance components	Proportion of variance
Situations	81%
Subjects	7%
Error	12%

	Reaction time in seconds		
	Situation 1	Situation 2	Situation 3
mean	9.79	5.07	7.21
sd	2.80	1.08	2.07

	Heart rate in beats per minute			
	Baseline	Situation 1	Situation 2	Situation 3
mean	79.35	88.72	92.57	84.28
sd	11.62	13.88	15.99	13.28

this that subjects differed in the effectiveness with which they coped with task demands. Similar results for the components of variance were observed when heart rate difference scores between task and baseline periods were calculated.

DISCUSSION AND CONCLUSIONS

The problem of quantifying experimental load was effectively resolved in both of the experiments described here. In each case it was shown that the independent variable (when manipulated along one dimension) controlled the dependent variable, and that moderating variables can be related to behavioural performance as indicators of emotional response and coping. It was possible to order the test conditions according to intensity, since they varied along a single dimension. If different test conditions cannot be ordered along a single dimension, we can only speculate about the effects on the dependent variable.

The problem of which parameter to take as the dependent variable was considered different in both experiments. The solution was first to look for a behavioural parameter which is closely related to the independent variables (levels of experimental load: presentation times in Experiment 1, performance in Experiment 2). In Experiment 2, the between-subject variance was lowest for percentage of failure (7%), intermediate for the dependent variable reaction time (31%), and higher still for the moderating variable, heart rate (87%). Hence, heart rate was controlled to a lesser extent by the independent variables, but reflected to a greater extent how different subjects coped with the tasks. In other circumstances, such as physical exercise, heart rate is related more closely to the experimental load, and cardiovascular behaviour is correspondingly more uniform among subjects.

A third advantage of this approach is that it enables precise analysis of the effects of order of presentation. If tasks cannot be regarded as levels along the same dimension, it becomes difficult to evaluate order, since different results may be due either to order or variation among task demands. By using the procedure administered in Experiment 1, we can conclude that different orders of presentation do lead to variable experimental demands, and hence different response patterns.

A final point concerns the duration of tests. The problem in many experiments is that the load characteristics change with prolonged administration, perhaps because subjects tire or stop working hard. The algorithm used in Experiment 2 goes some way towards resolving this problem, since the stimulus presentation is dependent in part on behaviour. It is possible to compensate for the changes in performance levels that occur over time, so that the overall pattern of performance (in percentage of failure) remains constant. In this way, subjects can be placed under constant load. It is likely that in the analysis of cardiovascular responses to mental stress tests, this factor is of major importance. The use of such procedures may enable us to come closer in the laboratory to modelling the chronic demands of everyday life.

REFERENCES

Abels D (1965) Konzentrations-Verlaufstest K-V-T. Hogrefe, Göttingen

Kuhmann W (1979) Realisation eines Verfahrens zur experimentellen Anforderungssteuerung bei einer einfachen Konzentrationsaufgabe. In: Bericht über das IV Kolloquium 'Psychophysiologische Methodik', 13, bis 16. Spitzingsee, Obb, p18-21

Lachnit H, Meinardus B (1981) Psychophysiologische Aspekte der Bewältigung einer individuell standardisierten mehrphasigne Blastungsfolge - eine explorative Analyse. Master thesis, Giessen

Walschburger P, Lachnit H, Meinardus B (1980) Anforderung und überforderung - Ein Ansatz zur Diagnostik von Belastungs-Beanspruchungsprozessen. Arch Psychol 133: 293-321

CHAPTER 5

The Importance of Cardiovascular Reactivity to Different Types of Stress for the Development of Hypertension

Wolfgang Schulte and August Wilhelm von Eiff

Medizinische Universitätsklinik, Sigmund-Freud-Strasse 25, D-5300 BONN 1, Federal Republic of Germany

SUMMARY

Cardiovascular reactivity to emotional stress has been investigated in a series of studies relating to the general hypothesis linking stress with essential hypertension. Increased blood pressure reactivity to mental stress (mental arithmetic in the presence of distracting noise) was observed in established hypertensives, and in prehypertensive states as well. Hyperreactivity in prehypertensive states did not, however, occur in response to the cold pressor test. The different haemodynamic patterns underlying responses to these tests may account for this discrepancy. It is concluded that blood pressure hyperreactivity in prehypertensive states will only be exhibited in response to mental stress tests that induce prominent cardiac reactivity. Pharmacological studies give support to the notion that increased sympathetic drive may be responsible for this cardiovascular reactivity. It is concluded that heightened blood pressure responses under emotional stress may be significant in the development of hypertension.

INTRODUCTION

It is well known that blood pressure (BP) increases in response to the stimuli that impinge during daily life. Intra-arterial blood pressure monitoring studies have demonstrated that wide variations in blood pressure occur throughout the day (Millar-Craig et al 1978). Apart from physical activity, stress-induced changes in mental arousal seem to be the most significant influences on these blood pressure variations (Conway et al 1984). Since it has been argued that stress plays a role in the pathogenesis of essential hypertension (von Eiff 1970), the variability of blood pressure may be altered in the development of the disorder. Heightened increases of blood pressure in stressful situations may occur, leading to greater fluctuations in pressure level.

Littler et al (1978) demonstrated that systolic blood pressure variability increases with pressure level, and Watson et al (1980) consider that pressure level is one of the main influences on variability. However, most studies comparing pressure variability in normotensives and hypertensives have only shown greater variability in the latter by assessing the absolute values of individual standard deviations; when variability is expressed in terms of the variation coefficient (standard deviation/mean), no differences between hypertensives and normotensives have been recorded (Pessina et al 1980; Mancia et al 1983). These results are not therefore completely in accord with the hypothesis that blood pressure regulation is altered during the development of hypertension. However, the problems of standardising assessments of blood pressure variability should be taken into account. Not only are there technical difficulties in recording, but randomly-occurring stressful events and their relationships with blood pressure reactions must be considered. The question of heightened blood pressure reactions in hypertensives studied under field conditions remains unresolved.

Exact quantification of stimuli and cardiovascular reactions may be possible only under laboratory conditions. Thus a growing number of investigators have concentrated on acute stress testing. But the results relating blood pressure reactivity to hypertension have been contradictory in the laboratory as well. Melville and Raftery (1981) have argued that the significance of heightened blood pressure reactions may be limited, since reactivity may be related to blood pressure level. Several factors may account for the contradictory and misleading results that have been reported, including different measurement techniques, different stimuli, and different or non-homogeneous study groups. A series of studies relating cardiovascular reactivity to hypertension have therefore been carried out in our laboratory.

BLOOD PRESSURE UNDER RESTING CONDITIONS

The primary hypothesis was that blood pressure level is determined by two factors: current mental and metabolic activity, and individual differences in the central tuning of blood pressure regulation. From this point of view, different states of mental activity should lead to different blood pressure values; this can be tested most simply by comparing casual and resting blood pressures. In addition, the behaviour of blood pressure during different rest or stress periods should vary between individuals. These patterns imply that under resting conditions, hypertensives and normotensives should differ in the manner in which blood pressure level changes.

Blood pressure measurements were therefore performed on 50 male subjects under two conditions: on admission to hospital (casual blood pressure), and under strict resting conditions (near-basal blood pressure). Subjects were assigned to normotensive, borderline hypertensive and hypertensive groups according to casual blood pressure, and then on the basis of near-basal blood pressure (Schulte et al 1978). As anticipated, the mean blood pressure for the whole group differed between measurement conditions. In addition, assignment to the three groups differed markedly between measurement conditions, and only 38% of the subjects were assigned to the same group in each procedure. This is illustrated in Fig 1. After basal blood pressure values were standardised by linear regression to a level comparable with casual blood pressure, new group allocations improved the fit by only 16% to 54%.

This study indicated that blood pressure level depends not only on measurement conditions, but on individual patterns of cardiovascular regulation. Individuals varied in the degree to which blood pressure differed between these two measurement conditions, and the reduction in blood pressure from casual to basal conditions was more pronounced in hypertensives. It was

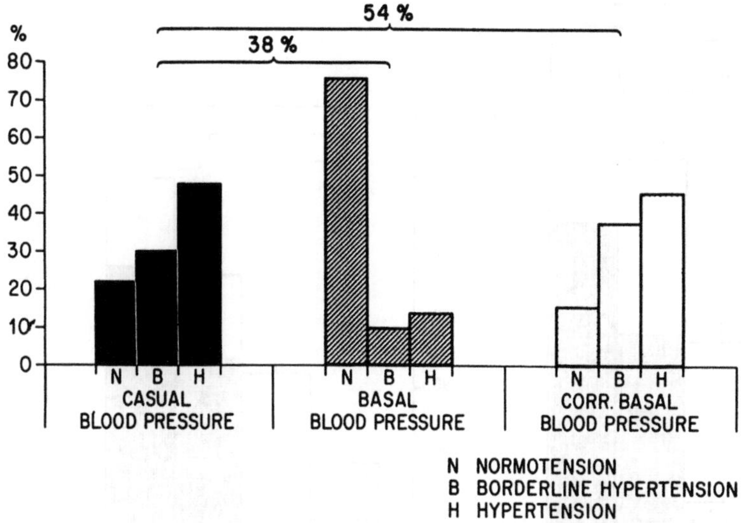

Fig 1. Percentage of 50 male subjects assigned to the three blood pressure groups on the basis of casual BP (left panel) and basal BP (centre panel). The right panel shows group allocation on basal BP, after this had been standardised to a level comparable with casual BP.

assumed by analogy that blood pressure reactions to stress might be stronger in hypertensives as well.

BLOOD PRESSURE REACTIVITY TO MENTAL STRESS

We have carried out several studies concerning blood pressure reactivity in hypertension. As a standard stimulus, mental arithmetic (addition of one- and two-digit numbers) during exposure to 90dB noise has been used. Blood pressure reactivity has been studied in different stages of hypertension and in borderline hypertension, while the influences of age and sex have also been considered (Schulte and Neus 1979; Schulte et al 1981a). The main result from these investigations is that hypertensives in Stage I and II, together with borderline hypertensives, show increased systolic and diastolic blood pressure reactivity in comparison with normotensive controls (see Fig 2). Hyperreactivity was demonstrated by assessment of both absolute and percentage changes. Hyperreactivity has been observed in both sexes, and is not age-dependent, although larger reactions are observed with increasing age.

The elevated blood pressure reactivity of established hypertensives may be secondary to vascular alterations. In this sense, it may be a phenomenon induced by hypertension, and not important in the initiation of the disorder. However, the fact that increased blood pressure reactivity is observed even in borderline hypertension cannot be due to hypertensive complications, but may result from a functional disorder.

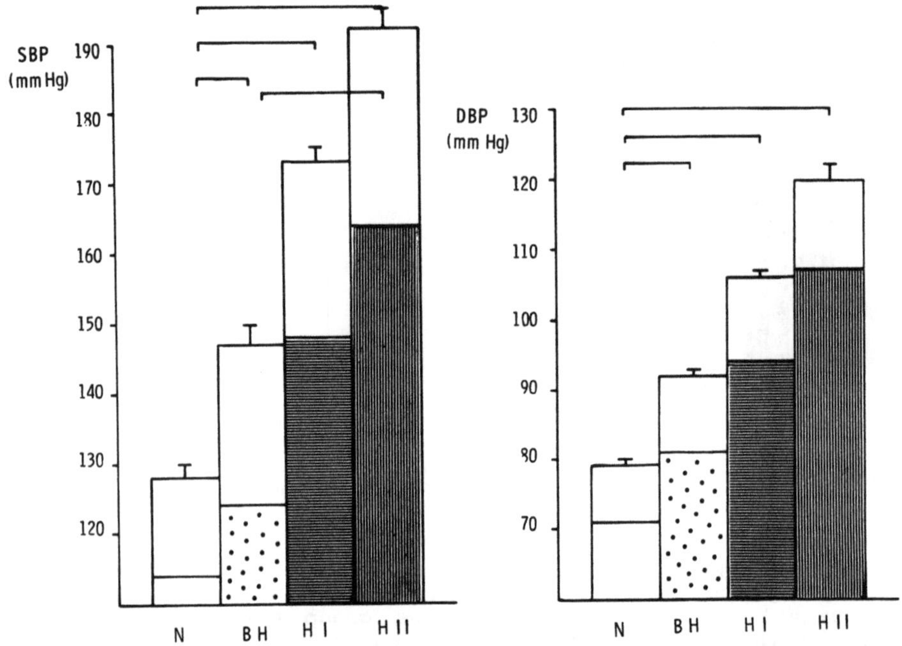

Fig 2. Resting values (hatched bars) and reactions (open bars) of systolic (SBP) and diastolic (DBP) blood pressure to mental stress in normotensives (N, n= 18), borderline hypertensives (BH, n= 17), hypertensives stage I (HI, n=30) and hypertensives Stage II (HII, n=13). Significant differences ($p<.05$) between the groups were tested by the Duncan test and are indicated by brackets.

Direct comparison of these results with other investigations that have involved mental arithmetic show some agreement (Shapiro et al 1963; Richter-Heinrich et al 1975). However, other authors have failed to confirm the effect (eg, Brod et al 1959; Sullivan et al 1981). Jern (1982) has also demonstrated blood pressure hyperreactivity in borderline hypertensives carrying out mental arithmetic. These contradictory results may be due to several factors, among which the resting value from which stress reactions are assessed is important. In our studies, the stress test is commenced only after a steady state of cardiovascular parameters is attained. Furthermore, the type of stimulus may be important. Blood pressure hyperreactivity among hypertensives has been described in response to visual puzzles and the Stroop test (Nestel 1969; Steptoe et al 1984). But as for mental arithmetic, other investigators have failed to observe different responses to these tasks in hypertensives and normotensives (Hollenberg et al 1981; Hjemdahl and Eliasson 1979).

In an attempt to resolve these discrepancies, we have studied the role of the type of stimulus in stress testing. A number of different stressors have been used, including noise (von Eiff et al 1982), the Type A structured interview (Rüddel et al 1983), ergometric exercise (Schulte et al 1983) and cold stimulation. The cold pressor test enjoys widespread use, so it was compared with the mental arithmetic test, and the results are summarised in Fig 3. As in the case of mental arithmetic, established hypertensives showed blood pressure hyperreactivity in comparison with normotensives, although the difference was significant only for systolic pressure. Borderline hypertensives did not differ from normotensive controls in their blood pressure responses.

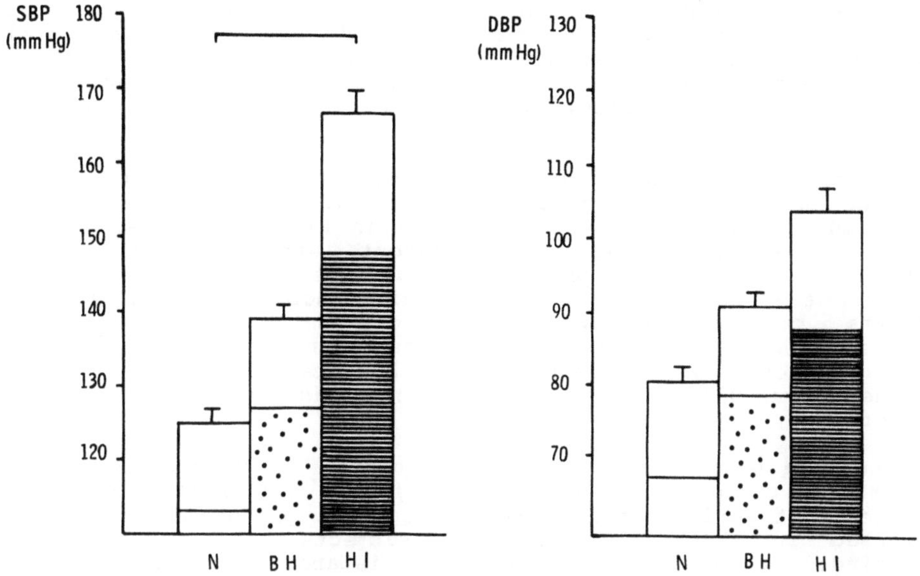

Fig 3. Resting values (hatched bars) and reactions (open bars) of systolic (SBP) and diastolic (DBP) blood pressure in the cold pressor test in normotensives (N), borderline hypertensives (BH) and hypertensives Stage I (HI). For other details, see Fig 2.

It is therefore suggested that general blood pressure hyper-reactivity independent of stimulus type may only exist in established hypertension. It is interesting that heightened pressure responses were not observed in the borderline state during cold stimulation. This indicates that in the early phases of the development of hypertension, blood pressure hyperreactivity depends on the type of stimulus. This result is consistent with the investigations of other authors. Heightened blood pressure responses to the cold pressor test have been recorded in several investigations of established hypertensives (Shapiro et al 1963; Greene et al 1965; Bühler et al 1983), but not in borderline hypertension (Robertson et al 1979).

HAEMODYNAMIC PATTERNS OF RESPONSE TO DIFFERENT STIMULI

The possibility that physiological differences in the pattern of haemodynamic change to different stimuli underlie variations in blood pressure behaviour was also studied. Cardiac output was determined by impedance cardiography in order to avoid disturbing stress reactions by invasive measurements, and total peripheral resistance was calculated in the usual way. The haemodynamic response to the mental stress test consisted of increases in blood pressure, cardiac output and heart rate. Mean stroke volume did not change, while total peripheral resistance showed a small but significant decrease. The increase of mean arterial pressure correlated with the change in cardiac output, but not with the change in peripheral resistance. The haemodynamic reaction of the same healthy subjects to the cold pressor test consisted of increases in blood pressure, cardiac output, heart rate and total peripheral resistance. Stroke volume fell slightly. In contrast with the mental stress test, the blood pressure increase correlated with peripheral resistance, but not with the change in cardiac output.

Direct comparison of haemodynamic responses to the two tests showed that the cardiac output response was stronger during mental stress than cold stimulation. As can be seen in Fig 4, the reaction in peripheral resistance was also different, with decreases in the mental stress test and increases during the cold pressor. These effects were evaluated in normotensive subjects, but similar patterns can be observed in borderline hypertensives and hypertensives. Thus during the mental stress test, the blood pressure reaction is determined mainly by a cardiac response, while peripheral vascular changes underlie the reaction during the cold pressor.

It should be noted that this result concerns the mean responses of the study groups. Further analysis of individual reactions revealed that during mental arithmetic, total peripheral resistance decreased in about two-thirds of subjects, and increased in one-third. Decreases in peripheral resistance were observed when the cardiac output response was greater than 12%. When the cardiac output response was smaller (or, in a few cases, negative), peripheral resistance increased. During the cold pressor test, cardiac output and peripheral resistance increased in almost all cases. The underlying mechanism for these different haemodynamic patterns may be that mental stress results in excitation of beta-receptors, while cold stimulation is predominantly associated with alpha-receptor excitation.

As far as comparisons between the groups are concerned, the hypertensives showed greater reactivity to mental stress than normotensives, not only in blood pressure, but also in heart rate and cardiac output. The haemodynamic responses to the cold pressor test showed few differences between groups, apart from a lower heart rate reaction from the hypertensive groups. The increase in total peripheral resistance tended to be stronger in established hypertensives as well.

These results suggest a physiological mechanism for the increased blood pressure reactivity of borderline hypertensives to mental stress but not cold stimulation. Blood pressure reactivity is only heightened in those tests that provoke predominantly cardiac reactions. It may be significant for the pathogenesis of hypertension that increased reactivity is only observed during beta-adrenergic cardiac stimulation.

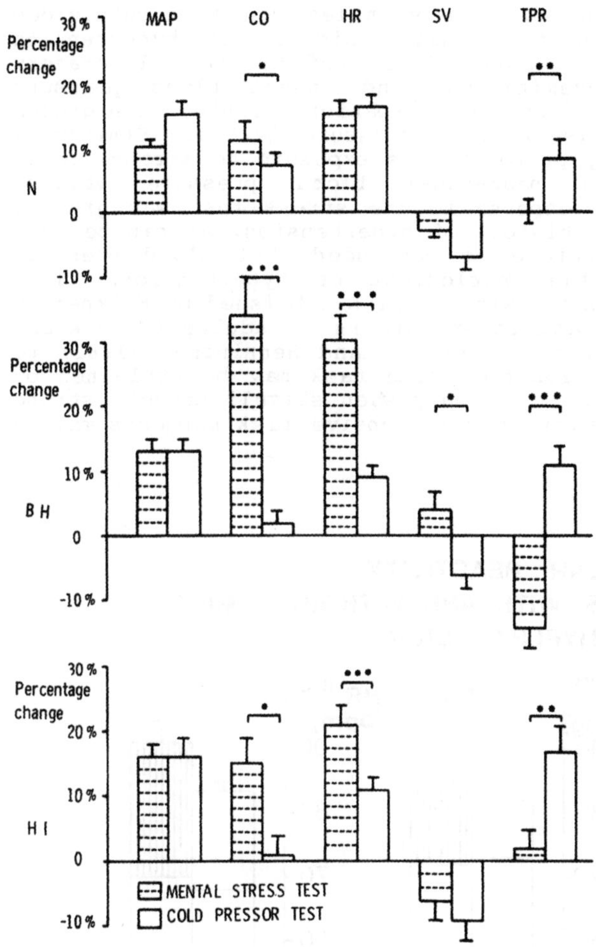

Fig 4. Comparison of the percentage reactions of mean arterial pressure (MAP), cardiac output (CO), heart rate (HR), stroke volume (SV) and total peripheral resistance (TPR) between the mental stress test and cold pressor test. N=normotensives, BH=borderline hypertensives, H=hypertensives. Significant differences *=$p<0.05$, **=$p<0.01$, ***=$p<0.001$.

BLOOD PRESSURE REACTIVITY IN PREHYPERTENSIVE STATES

It has already been noted that exaggerated blood pressure reactions in hypertension could be secondary to alterations in blood vessel geometry. However, similar patterns in borderline hypertension are more probably due to an autonomic imbalance, with heightened sympathetic stimulation of the heart. In order to buttress this hypothesis, cardiovascular reactions in prehypertensive states are of special interest. Two studies have been performed on normotensive 'risk' groups.

The first study involved 36 normotensive males, who were subdivided into those with and without a family history of hypertension (Schulte et al 1981b). Some days before the laboratory investigation, physical examinations and casual blood pressure measurement were carried out in the out-patient clinic. The groups did not differ in their casual blood pressure levels. Similarly, during the initial resting period before exposure to stress, there were no differences in near-basal blood pressure values. Nevertheless, the pressor response to the stress test was greater in the group with a family history of hypertension, as can be seen in Fig 5. It could therefore be concluded that blood pressure hyperreactivity precedes the development of hypertension. This point of view is in accordance with results published by Falkner et al (1979) on adolescents, and other authors on adults (Manuck and Proietti 1982). The failure of Ohlsson and Henningsen (1982) to find differences during a reaction time task may be explained by the type of stimulus used; it is only when stimuli provoke strong reactions that hyperreactivity in normotensive risk subjects may be apparent.

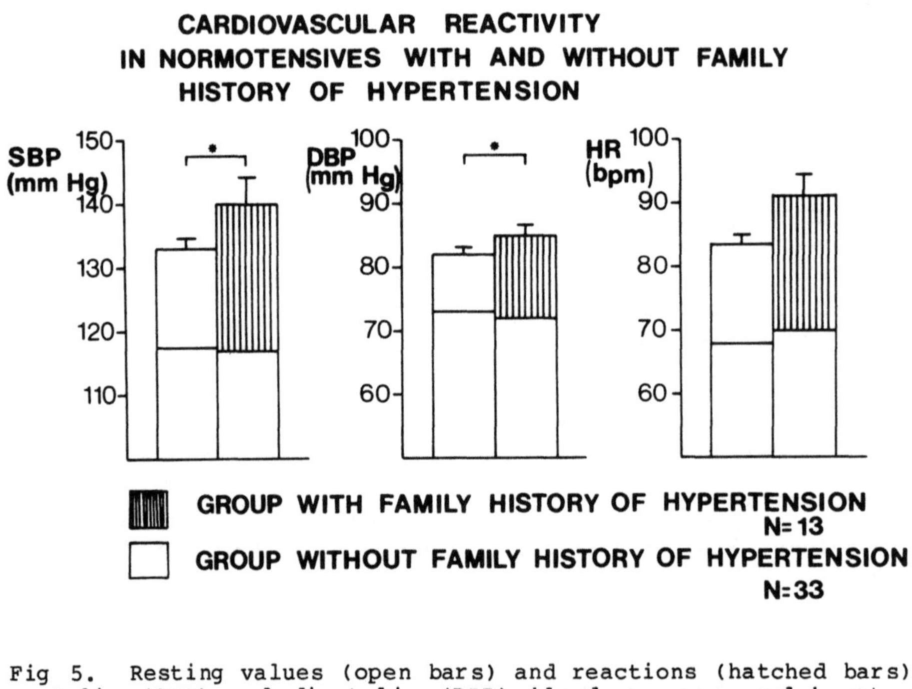

Fig 5. Resting values (open bars) and reactions (hatched bars) of systolic (SBP) and diastolic (DBP) blood pressure and heart rate (HR) in normotensives with and without a family history of hypertension. Significant differences are indicated by an asterisk (*=p<0.05).

A second approach to defining prehypertensive states involves studying normotensives with marginally elevated casual blood pressure (Schulte et al 1984a). A group of normotensive male subjects was divided into two subgroups according to seven casual blood pressure measurements. One group was defined as having one, or at most two, slightly elevated blood pressure readings, with all other values being normal. In the other group, all seven casual blood pressure readings were normal. The main results of this study was that the two groups did not differ in basal blood pressure, but those with marginally elevated casual blood pressure showed higher pressor responses to mental stress. Cardiac output and heart rate responses were also greater. However, during the cold pressor test the groups did not differ in their cardiovascular responses (see Fig 6).

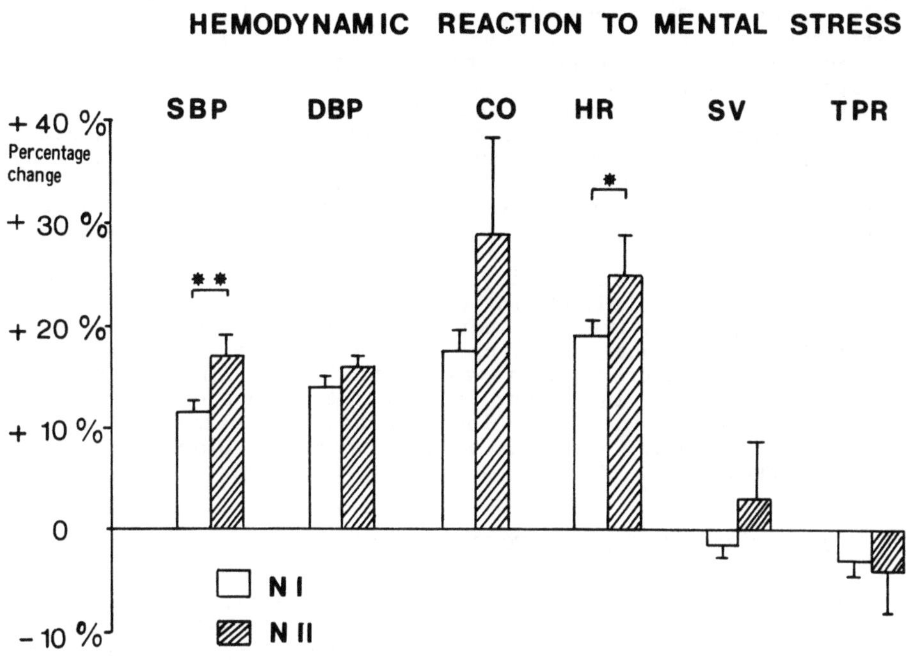

Fig 6. Haemodynamic reactions to mental stress in normotensives with marginally elevated casual blood pressure (NII, n=22) and control subjects (NI, n=46). For other details see Fig 4.

Both these studies reveal that the phenomenon of blood pressure hyperreactivity exists even before the onset of borderline or established hypertension. As in the studies of borderline hypertensives, blood pressure hyperreactivity was due to heightened cardiac reactions during mental stress.

INTEGRATED CARDIOVASCULAR RESPONSES TO STRESS

One conclusion that can be drawn from these studies is that cardiac hyperreactivity to mental stress represents an early phenomenon in the development of hypertension. Nevertheless, it should be noted that the peripheral blood vessels also participate in the increased blood pressure response. This is because when cardiac output shows a large increase, a large decrease in total peripheral resistance is to be expected. But since the changes in peripheral resistance did not differ between hypertensives and normotensives, the compensatory reduction in resistance among hypertensives was abnormally small. Thus it is suggested that the heightened blood pressure response to stress results from an integrated change in cardiovascular reactivity.

This assumption has been confirmed in a study examining the effects of beta-adrenergic blockade (Neus et al 1983). Hypertensives were treated either with beta-blockers (atenolol, 100mg p.d.) or diuretics (hydrochlorothiazide, 50mg p.d. + amiloride, 5mg p.d.) over six weeks). Cardiovascular reactivity to the mental stress test was then recorded. The results showed that the decrease in basal blood pressure produced by the medications was similar in the two groups. The blood pressure response to mental stress was also the same in the two groups, and did not differ from that of untreated hypertensives. Nevertheless, further analysis indicated that cardiovascular reactivity had been influenced by beta-blockade; in comparison with untreated hypertensives, the increase of cardiac output was smaller, while total peripheral resistance rose on average.

Similar patterns of reactivity during acute beta-adrenergic blockade have been obtained in normotensives by Bonelli et al (1979) and in hypertensives by Andrén and Hansson (1980). In each case, the cardiac output reaction to mental stress was diminished, while the blood pressure change was maintained. This observation supports the argument in favour of an integrated cardiovascular response to stress, in that blood pressure hyperreactivity is maintained by stronger vascular responses after blockade of the cardiac response.

THE ROLE OF THE SYMPATHETIC NERVOUS SYSTEM

These data suggest that blood pressure hyperreactivity may be due to augmented sympathetic stimulation of the circulation. But an alternative that should be taken into account is an increased sensitivity of beta-receptors. In order to test this possibility, the sensitivity of beta-receptors was determined by the isoproterenol test. The relationship between isoproterenol sensitivity and cardiovascular reactivity to mental stress was studied in a group of mild hypertensives. It was found that the CD25 (the dose which increased heart rate by 25bpm) correlated with systolic blood pressure and cardiac output increases during mental stress. The comparison with normotensives showed that the two groups did not differ in their CD25. This result indicates that the sensitivity of cardiac beta-receptors in inversely related to the degree of stimulation by mental stress. In addition, a super-sensitivity of beta-receptors in early hypertension could be excluded. Since blood pressure reactivity to mental stress was exaggerated in this hypertensive group, the effect was almost certainly due to stronger sympathetic activity.

Results from another investigation are consistent with this conclusion (Schulte et al 1984b). Methacholine was used to test sympathetic reactivity, since this substance induces a peripheral vasodilatation that leads to sympathetic counter-regulation with secondary increases of blood pressure and heart rate. Haemodynamic reactivity to methacholine was studied in a group of normotensives with genetic risk of hypertension, and a control group.

The results are summarised in Fig 7. The initial fall in blood pressure was the same in the two groups. During the phase of counter-regulation, the secondary rise in blood pressure was higher in the group with a family history of hypertension. Thus the haemodynamic response of this group could be attributed to higher sympathetic reactivity.

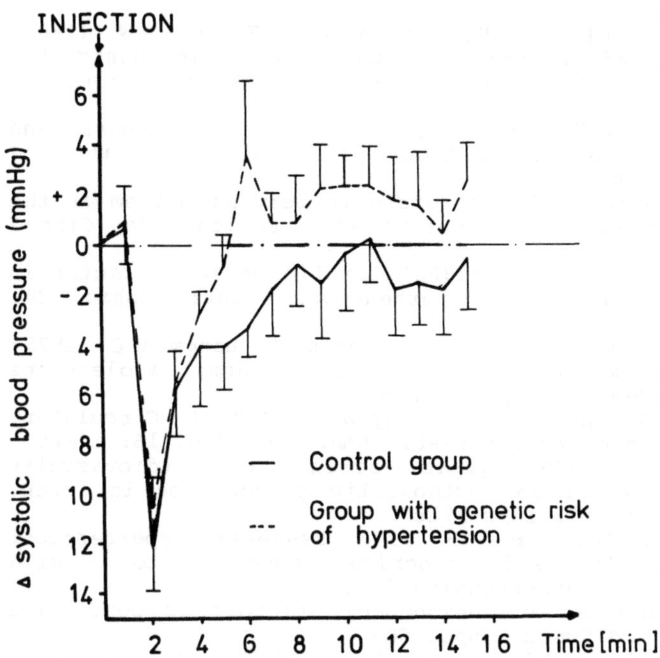

Fig 7. Course of systolic blood pressure reactions after injection of methacholine in normotensives with a genetic risk of hypertension (n=10) and a control group (n=8).

These results, in combination with the investigations of prehypertensive states described earlier, support the hypothesis that sympathetically-induced blood pressure reactivity is probably the earliest blood pressure phenomenon to be observed in the development of hypertension. Cross-sectional analyses of patients in different stages of hypertension indicate that blood pressure hyperreactivity to stress is a precursor of the disorder. However, this supposition must be proved by longitudinal study.

REFERENCES

Andrén L, Hansson L (1980) Circulatory effects of stress in essential hypertension. Acta Med Scand Suppl 646: 69-72

Bonelli J, Hörtnagl H, Brücke T, Magometschnigg D, Lochs H, Kaik G (1979) Effect of calculation stress on hemodynamics and plasma catecholamines before and after beta-blockade with propranolol (Inderal[R]) and mepindolol (Corindolan[R]). Europ J Clin Pharmacol 15: 1-8

Brod J, Fencl V, Hejl Z, Jirka J (1959) Circulatory changes underlying blood pressure elevation during acute emotional stress (mental arithmetic) in normotensive and hypertensive subjects. Clin Sci 18: 269-279

Bühler FR, Bolli P, Hulthén UL, Amann FW, Kiowski W (1983) Alpha-adrenoceptors, adrenaline and exaggerated vasoconstrictor response to stress in essential hypertension. Chest 83 (suppl): 304-306

Conway J, Boon N, Davies C, Jones JV, Sleight P (1984) Neural and humoral mechanisms involved in blood pressure variability. J Hypertension 2: 203-208

von Eiff AW (1970) The role of the autonomic nervous system in the etiology and pathogenesis of essential hypertension. Jap Circ J 34: 147-153

von Eiff AW, Friedrich G, Neus H (1982) Traffic noise, a factor in the pathogenesis of essential hypertension. Contr Nephrol 30: 82-86

Falkner B, Onesti G, Angelakos ET, Fernandes M, Langman C (1979) Cardiovascular response to mental stress in normal adolescents with hypertensive parents. Hypertension 1: 23-30

Greene MA, Boltax AJ, Lustig GA, Rogow E (1965) Circulatory dynamics during the cold pressor test. Amer J Cardiol 16: 54-60

Hjemdahl P, Eliasson K (1979) Sympatho-adrenal and cardiovascular response to mental stress and orthostatic provocation in latent hypertension. Clin Sci 57: 189s-191s

Hollenberg NK, Williams GH, Adams DF (1981) Essential hypertension: abnormal renal vascular and endocrine responses to a mild psychological stimulus. Hypertension 3: 11-17

Jern S (1982) Psychological and hemodynamic factors in borderline hypertension. Acta Med Scand: Suppl 662

Littler WA, West MJ, Honour AJ, Sleight P (1978) The variability of arterial pressure. Amer Heart J 95: 180-186

Lorimer AR, Macfarlane PW, Provan G, Duffy T, Lawrie TDV (1971) Blood pressure and catecholamine responses to 'stress' in normotensive and hypertensive subjects. Cardiovasc Res 5: 169-173

Mancia G, Ferrari A, Gregorini L, Parati G, Pomidossi G, Bertinieri G, Grassi G, di Rienzo M, Pedotti A, Zanchetti A (1983) Blood pressure and heart rate variability in normotensive and hypertensive human beings. Circ Res 53: 96-104

Manuck SB, Proietti JM (1982) Parental hypertension and cardiovascular response to cognitive and isometric challenge. Psychophysiology 18: 481-489

Melville DI, Raftery EB (1981) Blood pressure changes during acute mental stress in hypertensive subjects using the Oxford intra-arterial system. J Psychosom Res 25: 487-497

Millar-Craig MW, Bishop CN, Raftery EB (1978) Circadian variation of blood-pressure. Lancet 1: 795-797

Nestel PJ (1969) Blood-pressure and catecholamine excretion after mental stress in labile hypertension. Lancet 1: 692-694

Neus H, von Eiff AW, Friedrich G, Rüddel H, Schulte W (1983) Der Einfluss von Betarezeptorenblockern und Diuretika auf die Reaktivität des Blutdrucks. Z Kardiol 72: 617-621
Ohlsson O, Henningsen NC (1982) Blood pressure, cardiac output and systemic vascular resistance during rest, muscle work, cold pressure test and psychological stress. Acta Med Scand 212: 329-336
Pessina AC, Mormino P, Semplicini A, Palatini P, Casiglia E, Hlede M, Dal Palu C (1980) The blood pressure in 'labile' and 'established' hypertension: computer analysis of the continuous blood pressure recording. In: Stott FD, Raftery EB, Goulding L (eds) Proceedings of the Third International Symposium on Ambulatory Monitoring. Academic Press, London, p253-260
Richter-Heinrich E, Knust U, Müller W, Schmidt K-H, Sprung H (1975) Psychophysiological investigations in essential hypertensives. J Psychosom Res 19: 251-258
Robertson D, Shand DG, Hollifield JW, Nies AS, Frölich JC, Oates JA (1979) Alterations in the responses of the sympathetic nervous system and renin in borderline hypertension. Hypertension 1: 118-124
Rüddel H, Gogolin E, Friedrich G, Schulte W (1983) Coronary-prone behaviour and blood pressure reactivity in laboratory and life stress. In: Dembroski TM, Schmidt TH, Blümchen G (eds) Biobehavioral Bases of Coronary Heart Disease. Karger, Basel, p185-196
Schulte W, Neus H, Noffke H-U, von Eiff AW (1978) Zur Problematik der Einteilung in Blutdruckgruppen auf Grund von Ruhemessungen. Verh Dtsch Ges Inn Med 84: 789-792
Schulte W, Neus H (1979) Bedeutung von Stressreaktionen in der Hypertoniediagnostik. Herz/Kreislauf 11: 541-546
Schulte W, Neus H, von Eiff, A W (1981a) Blutdruckreaktivität unter emotionalem Stress bei unkomplizierten Formen des Hochdrucks. Klin Wochenschr 59: 1243-1249
Schulte W, Neus H, Rüddel H (1981b) Zum Blutdruckverhalten unter emotionalem Stress bei Normotonikern mit familiärer Hypertonieanamnese. Med Welt 32: 1135-1137
Schulte W, Fehring C, Neus H (1983) Cardiovascular reactivity to ergometric exercise in mild hypertension. Cardiology 70: 50-56
Schulte W, von Eiff AW, Neus H, Rüddel H (1984a) Cardiac hyperreactivity - an early phenomenon in the development of hypertension. Europ Heart J 5 (Abstr Suppl 1): 237
Schulte W, Diederich N, von Eiff AW, Neus H (1984b) Cardiovascular reactivity to methacholine in normotensives with genetic risk of hypertension. Clin Exp Hypertension 6: 717-730
Shapiro AP, Moutsos SE, Krifcher E (1963) Patterns of pressor response to noxious stimuli in normal, hypertensive and diabetic subjects. J Clin Invest 42: 1890-1898
Steptoe A, Melville D, Ross A (1984) Behavioral response demands, cardiovascular reactivity and essential hypertension. Psychosom Med 46: 33-48
Sullivan PA, Procci WR, de Quattro V, Schoentgen S, Levine D, van der Meulen J, Bornheimer JF (1981) Anger, anxiety, guilt and increased basal and stress-induced neurogenic tone: causes or effects in primary hypertension? Clin Sci 61: 389s-392s
Watson RDS, Stallard TJ, Flinn RM, Littler WA (1980) Factors determining direct arterial pressure and its variability in hypertensive man. Hypertension 2: 333-341

CHAPTER 6

Mental Stress Testing in the Physician's Office and in the Field

Heinz Rüddel, Hermann Neus, Wolf Langewitz, August Wilhelm von Eiff and Mark E McKinney

Medizinische Universitätsklinik, Sigmund-Freud-Strasse 25, D-5300 BONN 1, Federal Republic of Germany
and
Department of Preventive and Stress Medicine, University of Nebraska

SUMMARY

Simple mental stress tests were performed in the laboratory, the physician's office, and at the worksite in adults and children. The average blood pressure increases seen in children during a simple mental arithmetic task were similar in all three conditions. In hypertensive middle aged male executives, free of medication, similar blood pressure levels were attained during video task performance at the worksite and mental arithmetic in the laboratory. However, resting blood pressures were higher at the worksite. Hence, hyperreactivity to mental challenge in patients with mild hypertension was not observed during performance of a video game at the worksite. When mental stress testing in the physician's office was repeated, test-retest correlations of $r=.79$ for systolic blood pressure and $r=.87$ for diastolic blood pressure ($p<.001$) were recorded. These indicate that blood pressure reactivity is highly reproducible and might well be applied in the early diagnosis and the management of hypertension.

INTRODUCTION

It is well documented that mortality and morbidity from myocardial infarction increases with elevated clinical casual blood pressure, both diastolic and systolic (Kannel et al 1971). However, a strong relationship between blood pressure and mortality is only found at blood pressure levels higher than 100mm Hg diastolic. Furthermore, organ damage is better predicted by average blood pressure during the day than by casual blood pressure. Perloff et al (1983) showed prospectively that average blood pressure at work predicted later hypertension better than clinical casual blood pressure. In a forty-five year follow-up study of 142 subjects, Wood et al (1984) demonstrated the value of a positive blood pressure response to the cold pressor test as a predictor of later hypertension. The assessment of blood pressure reactivity to tasks evoking 'beta-adrenergic' responses permits observation of a pathophysiological mechanism that is probably relevant to the development of essential hypertension (Obrist 1981).

The finding of Falkner et al (1981) that blood pressure reactions to mental arithmetic predicted later hypertension in adolescents has been corroborated for adults. In a pilot study (von Eiff 1984), we observed a correlation between blood pressure responses to mental arithmetic and changes of casual blood pressure over two years ($r=.34$). In another study, blood pressure was monitored monthly in the afternoons using the Remler ambulatory system

(Langewitz et al, in press, a). Children who produced large increases in heart rate during mental arithmetic also showed progressively increasing average blood pressure. Those with a heart rate response below the group average did not increase in blood pressure. In this latter experiment, increase of heart rate was chosen as the discriminant variable because this response is much more stable in children than the blood pressure response.

Eliot has described a number of single cases with normal clinical casual pressure who showed an elevated and inappropriate cardiovascular response to mental challenge (increase of total systemic resistance during a testing procedure with predominantly cardiac stimulation), and subsequently experienced fatal cardiovascular events within a one-year follow-up (Eliot and Buell 1983). Different haemodynamic reaction patterns to mental stress in the laboratory can be consistently demonstrated in normotensives and patients with borderline, mild and severe hypertension (see Chapter 5). More important, subjects with a positive family history of hypertension exhibit similar haemodynamic reactions to those of hypertensives, providing further evidence that an exaggerated response to mental stress is a predictor of future elevations in casual blood pressure (Falkner et al 1979; Hastrup et al 1982). In 1976, von Eiff noted that patients with a personal history of high blood pressure had elevated reactions to mental challenge, even though they were normotensive when examined. These results suggest that it would be beneficial to perform a simple reactivity test during blood pressure screening, or in the physician's office, which does not need large and costly equipment. Only a very small number of patients can be tested in medical centres or laboratories. It is unlikely that 'normal', healthy people, without any serious complaints, will find their way into these institutions, so the results may not be representative.

A set of experiments was therefore carried out to determine whether blood pressure during mental challenge can be validly assessed in the field, whether hypertensives and normotensives still differ in blood pressure response when challenged outside the laboratory, and whether mental stress testing in the field is sensitive enough for the analyses of clinically relevant problems. We also assessed test-retest stability of the haemodynamic response in the physician's office. It has been possible to show that simple mental tasks lead to reproducible and stable increases in blood pressure and heart rate, and that the haemodynamic response to mental challenge is related to casual blood pressure as well as to average blood pressure during the day, as assessed by ambulatory blood pressure monitoring.

VALIDITY OF MENTAL STRESS TESTING IN THE FIELD

Under field conditions, a mental stress test paradigm has to be easy to apply. In our laboratory we have worked with a mental arithmetic plus noise test since the early 1960s (von Eiff 1976, see Chapters 5 and 7). Unfortunately, a slide projector, screen and tape recorder are necessary for the administration of the test, so its application in the physician's office or under field conditions is limited. We therefore compared a simpler mental arithmetic task with the standard mental arithmetic plus noise task in an A/B-B/A randomised design in healthy male students (Langewitz et al, in press, b). In the simple mental arithmetic task, subjects had to add up one-digit numbers played from a tape, onto which a counter-rhythmic beat from a metronome was also recorded (a 'click' every two seconds compared with the three-second interval between numbers). The task was presented with moderately challenging instructions. The blood pressure and heart rate reactions to the two tasks were nearly identical (13/8 vs 11/10mm Hg and 10 vs 11 bpm respectively). The resting baselines were 130±10/83±8 and 129±10/82±8 respectively, with an average heart rate of 71±6 and 71±6 bpm in both conditions.

Young boys aged 10-12 years were also examined in the laboratory or in the physician's office using the simple mental arithmetic task. No significant differences were found in blood pressure and heart rate response (9/9 vs 9/9mm Hg and 8 vs 7 bpm respectively). Resting blood pressures were 104±10/63±9 and 105±10/64±9mm, while the resting heart rates were 82±10 and 80±11 bpm in the laboratory and office. A larger sample of 180 children of the same age were subsequently tested with the simple mental arithmetic task in their schools (resting blood pressure averaged 108±10/76±9). Blood pressure increased 11±7/7±7mm and heart rate 18±14 bpm. Resting blood pressure was higher in the classroom than in the physician's office or laboratory, as was the increase in heart rate. However, the average blood pressure increases of 14/7mm during a simple arithmetic task in the classroom, 9/9mm in the physician's office and 9/9mm in the laboratory were similar. These results encouraged us to perform mental stress tests at the worksite in adults.

Departments of the Federal Government in Bonn were screened for people with elevated blood pressure. Eighty-nine civil servants who were free of any medication and had primary diagnosed mild hypertension (later confirmed as essential hypertension) were tested at the worksite, and two weeks later in the laboratory. The average of four clinical casual blood pressure values taken with a random zero sphygmomanometer was 153±13/97±9mm Hg. Blood pressure and heart rate were then measured every minute with an oscillometric automatic device (DATASCOPE) at the worksite during a resting period of three minutes and while playing a video game (Break Out^{R} ATARI). About two weeks later, all patients were retested in the laboratory at 10.00am. Here, blood pressure was measured with a two probe ultrasonic device every minute, while heart rate was assessed continuously by ECG. An initial adaptation period was followed by a resting period of ten minutes. Patients were then asked to perform a mental arithmetic task without speaking. One- and two-digit numbers were given over headphones together with interfering noise, and these had to be added up continuously. Two different mental stress tests were used to avoid adaptation effects.

Blood pressure reactivity to mental stress was calculated as the average blood pressure during either the video game (first two minutes) or the mental arithmetic task (second to fifth minute), minus the average resting values prior to the two tests (last two minutes in the field and five minutes in the laboratory). Absolute levels of blood pressure and heart rate did not differ significantly during the video game and the mental arithmetic task, as can be seen in Fig 1. Significant correlations between tasks were observed for systolic and diastolic pressure (r=.67 and .34 respectively), and for heart rate (r=.66). However, blood pressure and heart rate at rest were significantly higher at the worksite compared with the laboratory. Thus, reactivity to mental challenge

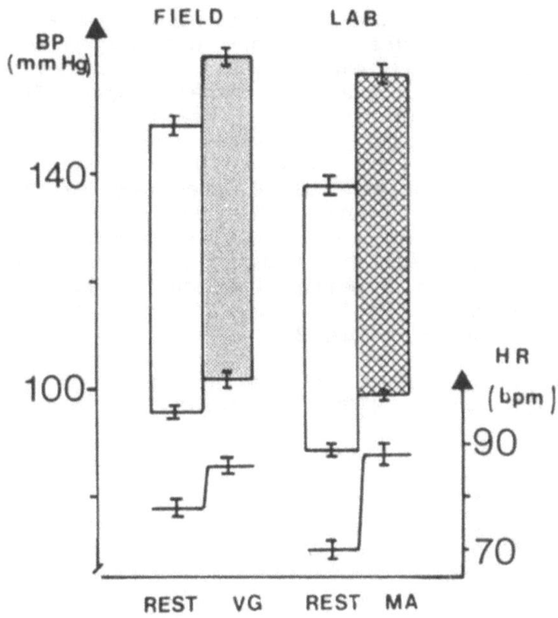

Fig 1. Blood pressure and heart rate before and during mental stress testing at worksite and in the hospital laboratory (VG = video game, MA = mental arithmetic) in patients with mild essential hypertension.

was different under the two conditions. In the field, blood pressure increased 13±8/6±6mm during the video game while increases of 21±10/10±6mm were recorded in the laboratory during mental arithmetic.

Previous experiments with middle-aged executives in Omaha, USA, revealed similar increases in blood pressure and heart rate during a mental arithmetic task (counting backwards by seven) and playing the Atari Break Out[R] video game. German middle-aged executives seem, however, to be less challenged by playing a video game or stress testing at the worksite compared to testing in the laboratory. Despite the different cardiovascular reactions elicited by the two mental stress test paradigms, heart rate and blood pressure level during both stress conditions were nearly identical and well correlated. This suggests that blood pressure during mental challenge is a reliable characteristic of patients under mental stress, even if the level of challenge and resting conditions are different, and that it can easily be assessed in field conditions. Further evidence that blood pressure reactions to simple stimuli that can be employed in the physician's office are useful was recently provided in the cold pressor study by Wood et al (1984) alluded to earlier. Shorter follow-up studies have produced inconsistent results concerning the predictive value of the cold pressor test.

During blood pressure screening, we also assessed blood pressure reactivity in a larger number of middle-aged male executives who did not undergo more detailed clinical evaluation. The subjects were classified as hypertensive (H), borderline hypertensive (BH) and normotensive (N) on one blood pressure reading during screening. Two hundred and ninety-eight men were classified as hypertensive, with a screening pressure above 160/95mm Hg. A further 248 were borderline hypertensive (blood pressure in the range 140-159/90-95mm), and there were thirty-six healthy normotensives. The three groups differed in age, so age-matching was carried out, and the following results are based on 36 men per group. Blood pressure was higher in hypertensive and borderline groups on screening, before and while playing the video game in the stress booth. These effects are shown in Fig 2.

Heart rate was also significantly higher in the hypertensive and borderline groups (means at rest: H=78±15, BH=75±10, N=70±8

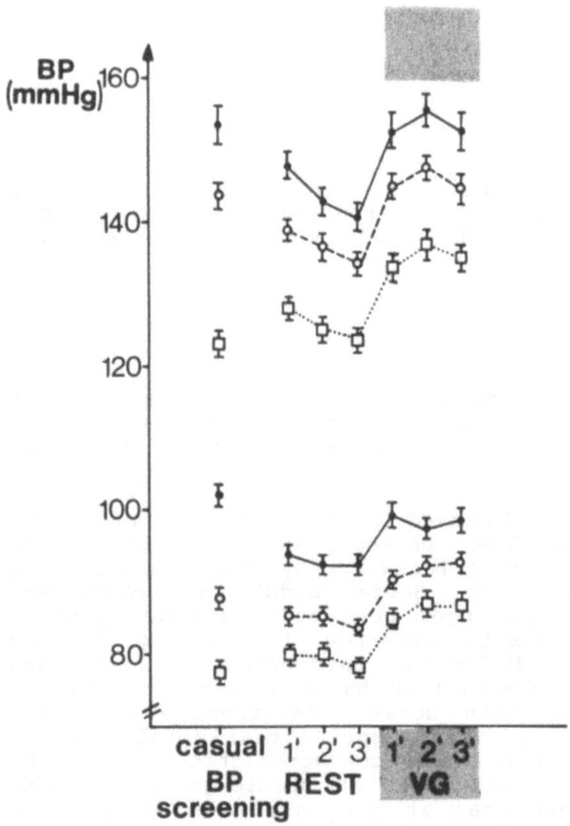

Fig 2. Casual blood pressure, and pressure before and during mental stress testing at worksite (VG=video game) in middle-aged civil servants (normotensives = □......□, borderline hypertensives = O------O, hypertensives = •——•)

bpm; means during the video game: H=89±14, BH=86±12, N=80±12). However, blood pressure and heart rate reactivity in the stress booth did not differ between the three groups. So when tested in the field, blood pressure hyperreactivity was not observed during the video game in hypertensives. On the other hand, when blood pressure levels during mental challenge are assessed, mental stress testing in the field yields reliable differences between groups, paralleling those observed in the laboratory.

It can be concluded that when baseline conditions are not well controlled, differences in reactivity may be hard to detect. Clinical casual blood pressure is often characterised simply as 'taken after sitting for five minutes'. When mental stress testing is performed, pretest resting conditions becomes crucial for the analysis of the results. 'Baseline values' in the field may be higher than those in the laboratory. In an earlier study, we manipulated baseline conditions and compared the average clinical casual blood pressure with the average resting blood pressure in the laboratory, and with the average pressure in the physician's office (Schulte et al 1979). In all three conditions blood pressure was read in a recumbent position, but subjects stood for five minutes before lying down (modified Meesman manoeuvre). This simple procedure decreased resting blood pressure in the physician's office significantly in comparison with blood pressure taken while lying down for 10 minutes. However, after the modified Meesman manoeuvre, diastolic pressure was still higher than resting blood pressure in the laboratory, while a difference in systolic pressure between resting in the laboratory and in the physician's office could no longer be observed. This method might well be useful in the assessment of basal levels. Obrist's (1981) suggestion of recording a true baseline the day after the experiment is practically impossible in many clinical or epidemiological studies.

SENSITIVITY OF MENTAL STRESS TESTING IN THE FIELD

In previous experiments, comparisons of blood pressure reactions to the mental arithmetic plus noise task in healthy normotensive males with and without a positive family history of hypertension have produced inconsistent results (Schulte et al 1981; von Eiff et al 1981). We hypothesised that the blood pressure increases were not large enough to yield significant hereditary effects. A methodological study on 49 male healthy students was therefore carried out, in which the challenge during mental arithmetic was increased. Nineteen of the students also carried out a simplified version in the physician's office. They were told to add up one- and two-digit numbers as fast and as accurately as possible. These numbers were presented in thirteen rows of fourteen numbers on a sheet of paper, as is shown in Table 1. The upper third of the sheet had a yellow background, and subjects were told that 85% of people could easily complete this section during a five-minute period; that the second, red, part could be finished by about 25%; and that only one person had ever finished adding up all the numbers in the third, blue, part correctly. Throughout the task, the time remaining for completing the calculations was given every minute. Emotionally disturbing noise (soccer game, accident, music, aircraft noise, crying children etc) was played between 80dB and 90dB (A) over earphones. In order to provide adequate resting data, the Meesman manoeuvre was used before the task.

Table 1. Mental arithmetic task (for details see text)

I	4 + 1 + 20 + 8 + 9 + 4 + 1 + 13 + 2 + 1 + 7 + 5 + 1 + 6 +
II	7 + 5 + 11 + 3 + 1 + 2 + 8 + 9 + 5 + 8 + 6 + 6 + 1 + 9 +
III	10 + 2 + 4 + 5 + 7 + 4 + 1 + 5 + 9 + 9 + 3 + 8 + 1 + 2 +
IV	1 + 6 + 2 + 4 + 15 + 7 + 1 + 6 + 4 + 17 + 1 + 3 + 4 + 5 +
V	4 + 9 + 5 + 2 + 8 + 3 + 4 + 5 + 8 + 7 + 10 + 9 + 5 + 7 +
VI	8 + 4 + 3 + 6 + 9 + 3 + 5 + 1 + 5 + 7 + 4 + 8 + 6 + 1 +
VII	7 + 3 + 12 + 7 + 4 + 8 + 9 + 4 + 3 + 5 + 6 + 8 + 5 + 8 +
VIII	14 + 1 + 7 + 8 + 2 + 5 + 6 + 7 + 9 + 4 + 9 + 6 + 2 + 1 +
IX	5 + 11 + 6 + 9 + 7 + 12 + 5 + 9 + 8 + 9 + 5 + 4 + 3 + 9 +
X	7 + 9 + 13 + 9 + 6 + 6 + 5 + 7 + 5 + 6 + 8 + 1 + 9 + 8 +
XI	12 + 6 + 8 + 7 + 14 + 9 + 8 + 7 + 7 + 7 + 9 + 1 + 10 +
XII	9 + 4 + 6 + 8 + 5 + 2 + 8 + 9 + 8 + 1 + 6 + 9 + 4 + 8 +
XIII	6 + 5 + 14 + 4 + 9 + 3 + 1 + 3 + 12 + 6 + 3 + 4 + 3 + 20 +

Subjects with a positive family history of hypertension (N=9) showed greater increases in blood pressure and heart rate in the laboratory and in the physician's office. The data collected in the physician's office are summarised in Fig 3. Mean resting blood pressures were $126\pm4/83\pm2$mm and $117\pm4/75\pm2$mm in the subjects with and without a positive family history. Resting heart rates were 76 ± 5 and 67 ± 5 bpm.

Systolic pressure reactions were significantly greater in the positive family history group, which heart rate increases were also greater in the first two minutes. No differences were observed in diastolic reactivity. We therefore conclude that haemodynamic differences in normotensive people will probably be detected if an adequate resting period is included in the experimental design, and the challenge is high enough to elicit an average systolic pressure response of at least 20mm Hg.

STABILITY OF BLOOD PRESSURE REACTIONS IN THE FIELD

High test-retest stability of blood pressure responses to mental challenge in the laboratory have been demonstrated by Manuck and Garland (1980) and others (for details see Chapters 2 and 8). Test-retest stability for the cold pressor test is also high (Wood

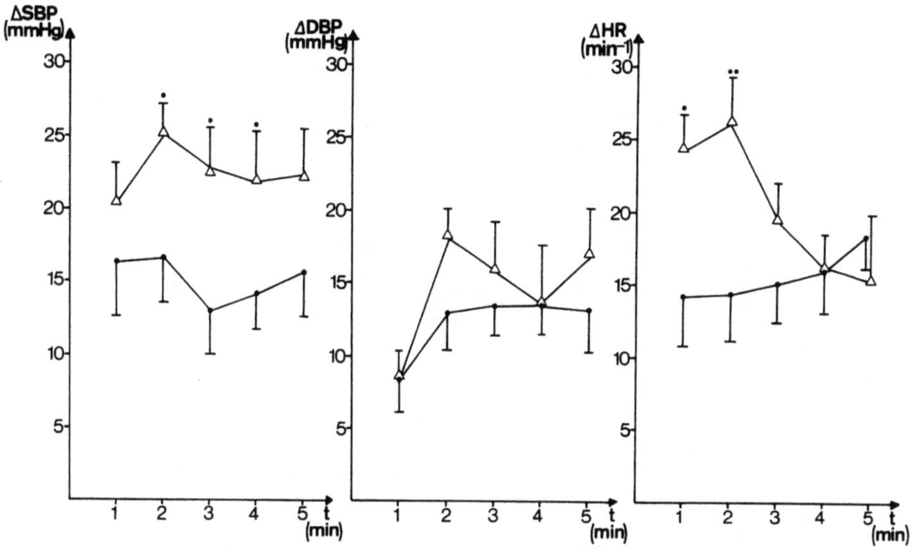

Fig 3. Blood pressure and heart rate changes during the modified mental arithmetic task in healthy normotensive male students with (Δ) and without (.) a positive family history of hypertension.

et al 1984). In a pilot study with healthy male executives in Omaha, USA, twenty-two middle-aged men (mean=47±7 years) performed mental stress testing and ambulatory blood pressure monitoring with the Avionics II system, and the procedure was repeated after six months (Rüddel et al, in press). Systolic and diastolic pressure ranges were similar on both days of ambulatory blood pressure monitoring. During the second day of ambulatory monitoring, diastolic pressure was lower than during the first (125±12/95±10mm compared with 125±12/86±8mm). Baseline and stress values in the laboratory did not differ between sessions. Test-retest correlations for average blood pressure readings were .52 for systolic pressure ($p<.05$) and .78 for diastolic pressure ($p<.01$). Test-retest correlations for resting blood pressure in the laboratory were .83 and .80 for diastolic pressure and .79 and .87 ($p <.001$) for average blood pressure during mental challenge. The original data set is given in Table 2.

A similar study was conducted on 57 healthy male executives (mean= 45±8 years) each of whom was tested twice within three months. Test-retest correlations for ambulatory blood pressure monitoring were .60 for systolic and .63 for diastolic pressure at work, while at home the corresponding correlation coefficients were .61 and .62. Test-retest correlations before and during mental challenge were: .56 and .72 (resting); .70 and .72 (video game); .62 and .56 (reaction time); and .62 and .66 (cold pressor test) (all $p<.01$).

These average blood pressure levels and test-retest correlations do not help predict if an individual patient will 'overreact' when retested several months later. A breakpoint analysis was therefore conducted, identifying all subjects with high standardised (z)

Table 2. Blood pressure during ambulatory monitoring and mental stress testing for each subject on two occasions

	Day One				Day Two			
Subject number	Average work BP SBP/DBP	BP range at work SBP/DBP	Baseline BP in the Lab SBP/DBP	Stress BP SBP/DBP	Average work BP SBP/DBP	BP range at work SBP/DBP	Baseline BP in the Lab SBP/DBP	Stress BP SBP/DBP
1	119/95	44/26	122/74	144/78	122/75	24/19	121/731	142/76
2	148/110	49/36	140/79	165/102	134/100	41/42	141/90	164/104
3	130/96	42/24	128/75	149/90	134/89	44/27	134/87	154/94
4	125/101	65/29	147/88	166/93	112/87	25/21	137/86	173/92
5	115/95	25/15	132/88	152/94	130/95	31/27	130/88	142/88
6	116/82	29/32	99/66	115/74	115/77	32/26	111/69	122/76
7	121/97	27/20	119/78	137/83	127/87	45/39	119/76	139/86
8	126/91	36/23	130/71	150/84	121/79	60/35	110/67	131/74
9	128/82	45/25	130/68	138/80	133/75	36/34	123/64	142/75
10	105/82	58/10	103/61	146/71	105/74	50/21	111/61	141/70
11	146/107	23/29	136/81	154/89	164/93	44/25	127/76	141/84
12	116/87	43/12	131/79	155/79	122/82	32/44	127/76	156/82
13	144/107	57/32	136/77	156/87	132/97	46/27	128/72	149/86
14	113/90	38/24	130/77	130/81	118/88	34/30	133/85	146/83
15	139/114	44/42	117/812	130/92	114/94	22/17	106/76	125/88
16	107/84	33/28	121/78	129/79	123/80	33/30	123/71	125/77
17	130/101	40/20	141/90	156/88	123/94	33/27	135/89	144/82
18	128/95	40/23	126/73	150/90	122/83	34/60	130/70	146/82
19	129/93	46/38	122/77	140/86	114/89	25/23	123/80	139/81
20	113/87	21/8	129/85	147/84	132/91	26/36	129/83	141/85
X±SD	125+12/95+10	40/25	127+12/77+7	145+13/85+7	125+12/86+8	36/31	125+9/77+9	143+13/83+8

SBP = Systolic blood pressure
DBP = Diastolic blood pressure

scores in systolic and diastolic pressure for any of the mental stress tasks. Seven out of eight of these individuals with high Z scores showed the same pattern of response during the second examination. We conclude that the test-retest stability of mental stress testing is high enough to justify use in ongoing epidemiological studies concerning risk factors for coronary heart disease and hypertension. It is no surprise if relatively poor test-retest correlations are reported for haemodynamic reactions to mental stress, since the error variance of change scores is twice that of the absolute values, when error variance at rest and during stress do not correlate. Although percentage changes from baseline are often reported in the physiologic literature, reporting absolute values at rest and absolute changes during mental challenge seems to be appropriate when describing clinical data on test-retest stability (see also Chapter 9).

During the last decade, increasing interest has been taken in ambulatory blood pressure monitoring, mainly because it has been assumed that blood pressure recorded during everyday activities should best describe the 'typical' pressure level and characteristic pressure variations of each patient. Indeed, during ambulatory blood pressure monitoring, the influence of sleep, physical activity, the circadian rhythm and self-reported arousal or behaviour can be documented (see Sokolow et al 1970; Pickering et al 1982). However, no two days in any person's life are identical, and therefore the range of individual blood pressure changes is large. This is reflected in the inconsistent results of test-retest reliability studies using non-invasive ambulatory blood pressure monitoring techniques (Des Combes et al 1984; Neus et al 1984).

Non-invasive systems probably record a relatively high number of incorrect readings (about 10%). The use of intra-arterial devices must be restricted to research settings because of the medical risks (eg, Littler 1976). Although it has been shown in some prospective studies that ambulatory pressure is a better predictor than casual blood pressure of later morbidity, concerns about technical difficulties, costs, and unsolved methodological problems remain. Recommendations about the general application of ambulatory techniques may only be made very cautiously. Mental stress testing in field conditions may provide a useful additional or alternative method of assessing individual blood pressure fluctuations and reactions to challenge. This, however, can only be proved in prospective studies that include both mental stress testing and ambulatory blood pressure monitoring. Very simple tasks like the mental arithmetic task performed in the presence of a distracting noise may elicit valid and reliable increases in blood pressure and heart rate in the physician's office, especially if care is taken to collect adequate baseline measures.

REFERENCES

Des Combes BJ, Porchet M, Waeber B, Brunner HR (1984) Ambulatory blood pressure recordings. Reproducibility and unpredictability. Hypertension 6: 100-114

von Eiff (1976) Die Dianose des Stress. In: von Eiff AW (ed) Seelische und körperliche Stoörungen durch Stress. Gustav Fischer Verlag, Stuttgart, New York, p194

von Eiff AW (1984) Pathophysiologie und Äthiologie der essentiellen Hypertonie. Münch Med Wochenschr 126: 165-170 von Eiff AW, Neus H, Friedrich G et al (1981) Festellung der erheblichen Belastung durch Verkehrslärm mit Mitteln der Stressforschung. Umweltbundesamt, Berlin

Eliot RS, Buell JC (1983) The role of the CNS in cardiovascular disorders. Hospital Practice 18: 189-199

Falkner B, Kushner H, Onesti G, Angelakos ET (1981) Cardiovascular characteristics of adolescents who develop essential hypertension. Hypertension 3: 521-527

Hastrup JL, Light KC, Obrist PA (1982) Parental hypertension and cardiovascular response to stress in healthy young adults. Psychophysiology 19: 615-622

Kannel WB, Gordan T, Schwartz MJ (1971) Systolic versus diastolic blood pressure and risk of coronary disease. Amer J Cardiol 27: 335-342

Langewitz W, von Eiff AW, Gogolin E, Neus H, Rüddel H, Schmieder R (in press, a) Reliability and validity of ambulatory blood pressure recording in children. Clin Exper Hypert

Langewitz W, Gogolin E, Neus H, Rüddel H (in press, b) Comparison of cardiovascular reactivity to mental arithmetic in a field and laboratory setting. Biol Psychol

Littler WA (1976) Median nerve palsy - complication of brachial artery cannulation. Postgrad Med J 52 (Suppl 7): 110-113

Manuck SB, Garland EN (1980) Stability of individual differences in cardiovascular reactivity: a thirteen-month follow-up. Physiol Behav 24: 621-624

Neus H, Gogolin E, Langewitz W, von Eiff AW (1984) Intermittent ambulatory blood pressure recordings in children. Methodological aspects and influence of family history on hypertension. Klin Wochenschr 62: 1038-1043

Obrist PA (1981) Cardiovascular Psychophysiology: a perspective. Plenum Press, New York

Perloff D, Sokolow M, Cowan R (1983) The prognostic value of ambulatory blood pressure. J Amer Med Ass 249: 2792-2798

Pickering TG, Harshfield GA, Kleinert HD, Blank S, Laragh JH (1982) Blood pressure during daily activities, sleep and exercise comparison of values in normotensive and hypertensive subjects. J Amer Med Ass 247: 992-996

Rüddel H, McKinney ME, Buell JC, Eliot RS (1984) Reliabilität es Cold Pressor Tests. Herzmedizin 7: 39-43

Rüddel H, McKinney ME, Dembroski TM, MacDougall JC, Grant LE, Witte H, Buell JC, Eliot RS (in press) Reliability of ambulatory blood pressure monitoring and blood pressure response to mental challenge in the laboratory. J Autonom Nerv System

Schulte W, Kersting B, Neus H (1979) Verhalten und Bestimmung des Ruheblutdrucks. Therapiewoche 29: 5764-5777

Schulte W, Neus H, Rüddel H (1981) Zum Butdruckverhalten unter emotionalem Stress bei Normotonikern mit familiaren Hypertonieanamnese. Med Welt 32: 1135-1137

Sokolow M, Werdegar D, Perloff D, Cowan RM, Brennenstuhl H (1970) Preliminary studies relating portable recorded blood pressure to daily life events in patients with essential hypertension. Bibli Psychiat 144: 164-189

Wood DL, Sheps SG, Elveback LR, Schirger A (1984) Cold pressor test as a predictor of hypertension. Hypertension 6: 301-306

Section II

Methodology of Stress Test Administration and Analysis

CHAPTER 7

Selected Topics in the Methodology of Stress Testing: Time Course, Gender and Adaptation

Hermann Neus and August Wilhelm von Eiff

Medizinische Universitätsklinik, Sigmund-Freud-Strasse 25, D-5300 BONN 1, Federal Republic of Germany

SUMMARY

The time course of haemodynamic reactions to mental stress has not received much attention in the psychophysiological literature. However, the data suggest that the time course of cardiovascular adaptation to active coping tasks might be important to the analysis of cardiovascular control mechanisms. With mental arithmetic, a steady state of circulatory parameters is reached after one to two minutes of stress exposure, so that representative stress values can be calculated. In the first two minutes transient changes occur. Evidence is presented that the arterial baroreflex and autoregulation contribute to these changes. The initial peak generally seen in heart rate seems to be indicative of sympathetic drive to the heart and functional hyperreactivity.

Female sex hormones may interfere with the autoregulatory adjustments typically found in men. The role of gender and female sex hormones must be considered in the analysis of experimental stress studies. Sex differences are linked both to hormonal and socioeconomic factors. When women participate in psychophysiological studies, the use of oral contraceptives, menstrual cycle phase and the menopause should be monitored.

In prospective and intervention studies, adaptation phenomena impose severe methodological problems. The pattern of reduced reactivity on repeated testing is particularly prominent in initially hyperreactive subjects.

THE RELEVANCE OF TIME COURSE IN STRESS TESTING

The time course of haemodynamic reactions to mental stress has not received adequate attention. In the present chapter we will try to show that investigation of the time course may be helpful to psychophysiological investigations, for a variety of reasons.

The complexity of cardiovascular adaptation during acute mental stress is one consequence of manifold physiological mechanisms controlling the circulation. Indeed, regulation of blood pressure is the net result of a multiple interaction of different control systems (Guyton 1980). The primary factors which have to be considered in cardiovascular reactivity to mental stress include neural and hormonal effects, such as the release of epinephrine and cortisol (Frankenhaeuser 1983). These primary processes provoke haemodynamic changes which in turn trigger secondary events. These include both arterial baroreceptor and low pressure receptor responses, since central venous pressure is reported to increase during mental stress (Brod et al 1979). Furthermore, when cardiac output (and tissue perfusion) is increased without a concomitant increase in metabolism, autoregulatory phenomena also come into play (Guyton et al 1973). The situation may be even more complicated in that mental stress not only affects haemodynamic variables directly, but also may influence these feedback mechanisms. In humans, this has been shown in the case of the arterial baroreflex by Conway and others (1983); recent experiments by Forsman and Lindblad (1983) suggest that afferents from the aortic receptors which regulate heart rate (see Mancia et al 1978) are most important under these circumstances, because responses were less obvious when carotid sinus baroreflex was selectively studied by neck suction. All these processes operate at different time constants. In view of this complex pattern, it is apparent that multiphasic responses are likely to occur.

Some additional prefatory remarks are in order, before the results of studies on time-course are detailed. Firstly, the phasic responses that occur on a beat-by-beat basis (for example in reaction to single noise bursts) will not be discussed here. Rather, analysis is restricted to cardiovascular reactions that occur over longer time periods, ranging from one to 30 minutes. This limitation has been imposed because there is evidence that reactions over this longer time scale are related to cardiovascular disease, and that heightened reactivity to mental stressors is observed in manifest (Schulte et al 1981) and in borderline hypertensives (Schulte and Neus 1979), as well as in individuals at genetic risk for developing the disorder (Falkner et al 1979). Short-term phasic responses are discussed in greater detail by Somsen (Chapter 13).

Similarly, discussion is confined to conditions that evoke a reactivity pattern reminiscent of ß-adrenergic stimulation. Typical examples include mental arithmetic, the Stroop test, video games, shock avoidance and reaction time tasks. The common psychological dimension of these stressors is that they require 'active coping'. The physiological reaction to such stressors is characterised by increases in heart rate and - at least for mental arithmetic - cardiac output (Brod et al 1959; Schulte and Neus 1983). This is different from other stressors, such as the cold pressor test, interviews (Rüddel et al, in press) or noise exposure (Neus and von Eiff in press; see also Chapters 1 and 5).

The principal variable measured in these studies is blood pressure. Usually it has been measured indirectly by auscultation on a minute-by-minute basis. Measurement errors therefore arise that have to be taken into account. Such errors can be attenuated by averaging across several readings; however, this averaging process is useful only when restricted to a steady state phase. Consequently, it is crucial to know whether, and when, a steady state of cardiovascular parameters is achieved during acute mental stress testing.

TIME COURSE OF BLOOD PRESSURE AND HEART RATE RESPONSES

In our laboratory we have used a standardized mental arithmetic task since 1969 (von Eiff et al 1969). Only minor changes have been imposed since that time. The general procedure is as follows: after a baseline (defined by a steady state of the measured circulatory parameters) is achieved, the mental arithmetic task is presented as a slide on a screen. The subject is instructed to add continuously one- and two-digit numbers for five minutes, as fast and as accurately as possible. Simultaneously, the subject is disturbed by affective noise of different types (for example jazz music, aircraft noise, soccer game, a quarrel, etc).

Some experiments have been performed with invasive measurement of blood pressure (Schirmer 1981). In Fig 1 an example of an individual reaction pattern is shown. Systolic and diastolic blood pressure increase after about 60 seconds and reach a steady state after 90 seconds. Heart rate rises immediately, reaches a peak after about 60 seconds and a steady state after 90 seconds. Note that both heart rate and blood pressure reach a steady state and that this occurs at the same time. In the phase when blood pressure rises there is a simultaneous decrease in heart rate. Similar results are also found in studies when blood pressure is measured indirectly every minute and heart rate is averaged across every minute. This is illustrated in Fig 2. During the first minute, systolic and diastolic blood pressure rise while heart rate exhibits its typical peak. From the second minute on, these parameters do not change appreciably. Analogous results have been found almost invariably in studies with this task. We therefore routinely consider that an average from the second to the fifth minute of stress exposure is representative, and determine reactivity on the basis of this interval, omitting the values measured during the first minute.

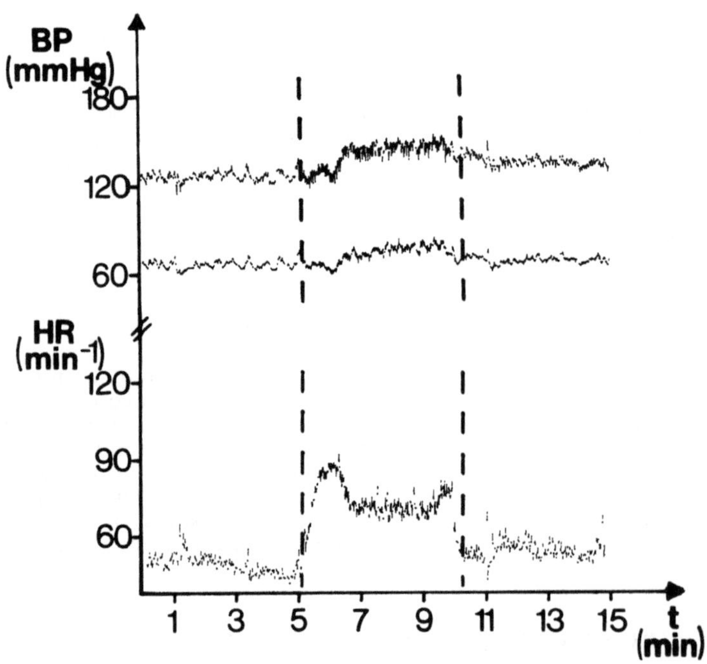

Fig 1. Example of an individual reaction to mental stress (top: systolic BP, middle: diastolic BP, bottom: HR). The stress period is indicated by vertical lines. (This picture was kindly provided by Dr G Schirmer.)

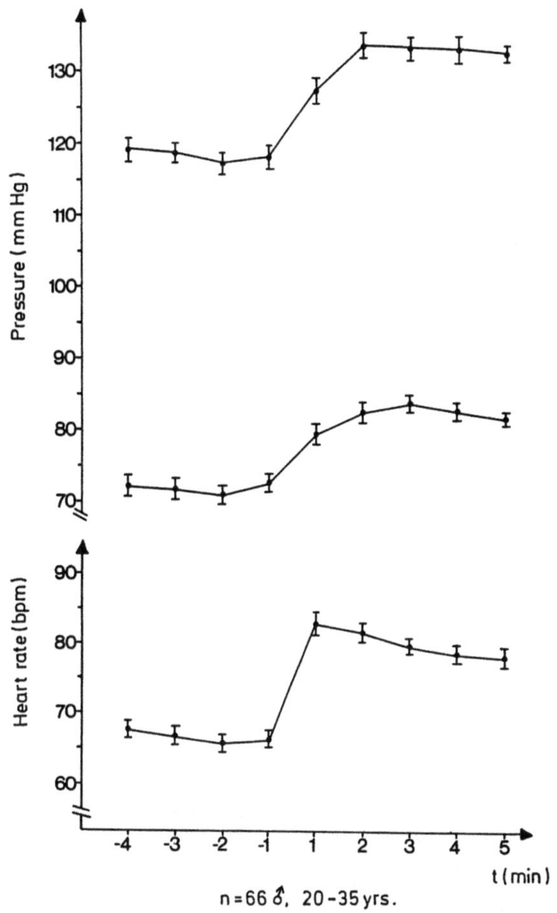

Fig 2. Average reaction of systolic BP (top), diastolic BP (middle) and HR (bottom) in a sample of 66 young healthy male subjects (mean ± SEM). Prestress period is indicated by negative, and stress period by positive time-markers.

The time course of heart rate in the experiment described above suggests that baroreceptor influences are operative during the stress period. This is indeed confirmed by recent experiments by Conway et al (1983) and Forsman and Lindblad (1983), although baroreflex sensitivity as determined by the Oxford method is diminished under stress. This led us to hypothesise that an impaired baroreflex sensitivity with respect to heart rate might contribute to the hyperreactivity of heart rate in essential hypertension (Bristow et al 1969; Korner et al 1974). It was expected that the secondary decrease in heart rate might be attenuated in hypertensives. This assumption was confirmed in the study outlined in Fig 3. The average reactions of heart rate in 15 young subjects with mild hypertension and 21 normotensive subjects of the same age are shown. While both groups exhibited the same initial peak in heart rate, the secondary decrease in heart rate was smaller in the hypertensive group (p = 0.027).

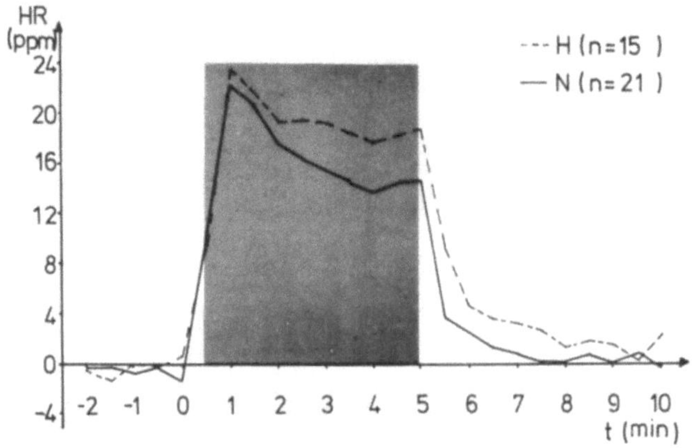

Fig 3. Time course of HR in a sample of 15 young mild hypertensives (H) and 21 normotensives (N) of the same age. HR was averaged every 30 sec. The stress period is indicated by the shaded area.

In contrast, the cardiovascular reactions of normotensive subjects with a positive family history of hypertension differ from this pattern. In Fig 4, the average reactions of ten subjects with and 20 subjects without a positive family history are summarised. These data were taken from a recent study from our laboratory in which we intensified the classical test by feedback of performance and time. Subjects with a positive family history differed from the control group in all variables. As to the heart rate reaction, it should be noted that the initial peak is enhanced in these subjects. This difference is maintained during the entire stress period, suggesting that the secondary decrease (mediated by the baroreflex) is the same in both groups. These data indicate that consideration of the time course of reactions might help to differentiate a 'functional' hyperreactivity from a 'structural' hyperreactivity which is secondary to vessel damage.

Unfortunately the time course of cardiovascular reactions is reported only in a minority of studies. In these cases an initial peak in heart rate has also been found by investigators using similar stressors (Falkner et al 1979; De Good 1975; Obrist et al 1978; Melville and Raftery 1981). In the experiments of Obrist et al (1978) using a reaction time task, an initial peak was also found for carotid dp/dt. The initial peaks of heart rate and carotid dp/dt disappeared after ß-blockade, indicating that they were due to sympathetic stimulation. Similar results were obtained in a recent study of the time course of heart rate responses to mental stress in two groups of male hypertensives who were treated either with a ß-blocker or with a diuretic (Neus et al 1983). In

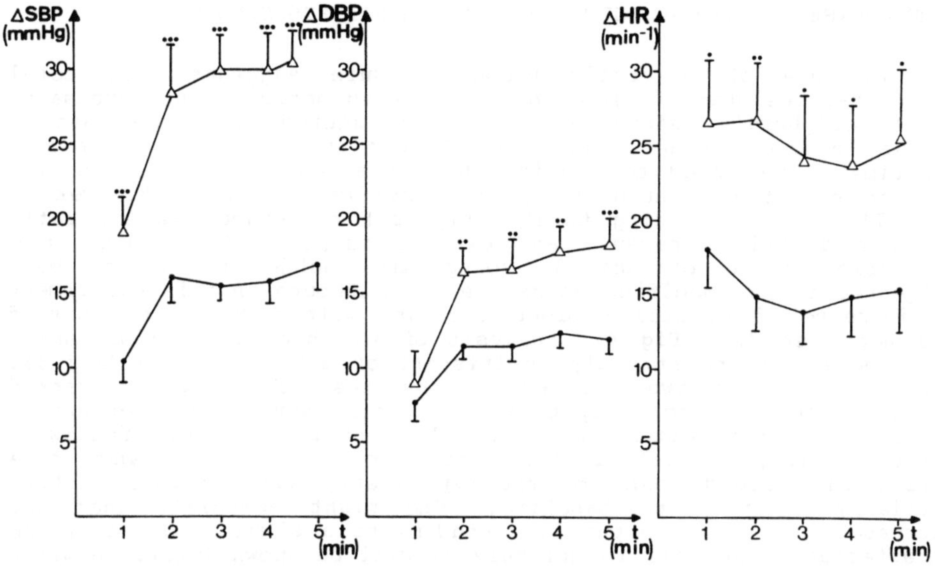

Fig 4. Average cardiovascular reactions of 10 young normotensive subjects with a family history of hypertension (triangles) and 20 control subjects without a family history (closed circles) (mean ±SEM). Points denote group differences (. $p<0.05$, .. $p<0.01$, ... $p<0.001$).

patients treated with diuretics there was a slight but significant decrease of heart rate from the first minute to the second and third minutes of stress exposure. On the other hand, patients treated with ß-blockers did not exhibit a peak in heart rate response, but maintained smaller stable responses throughout the task. However, blood pressure results different from ours have been reported. Some investigators observe peaks in systolic blood pressure similar to those in heart rate (De Good 1975; Obrist et al 1978; Melville and Raftery 1981), but except for one study (Nyberg 1976), stressors other than mental arithmetic were used (shock avoidance, reaction time tasks, Stroop test). Falkner et al (1979), using mental arithmetic, did not find a peak in systolic blood pressure. Thus, the time course of systolic blood pressure might depend on the stressor used. This suggests that it may be necessary to differentiate among the class of the so-called 'ß-adrenergic' stressors. Alternatively, other factors such as the subject pool or details in the experimental procedures might contribute to the differences in results that have emerged. In the case of diastolic blood pressure, a continuous increase during the stress period is apparent in some experiments, especially when longer stress periods (up to 15 minutes) are employed (De Good 1975; Obrist et al 1978). This increase is less obvious for the Stroop test or mental arithmetic of similar duration (Nyberg 1976; Falkner et al 1979; Melville and Raftery 1981).

TIME COURSE OF CARDIAC OUTPUT AND PERIPHERAL RESISTANCE

The increase of diastolic blood pressure suggests that total peripheral resistance rises over the stress period. An increase of total peripheral resistance might be triggered by an increase of cardiac output via autoregulation (Guyton et al 1973). In recent experiments we tried to examine this question directly, measuring cardiac output continuously by impedance cardiography (Kubicek et al 1974) and calculating total peripheral resistance as the ratio of mean arterial pressure and cardiac output. These data were available in 28 of the 30 young male subjects who had been subjected to the modified stress test. Although in this experiment an increase of diastolic blood pressure within the stress period was not seen (see Fig 4), analysis of the haemodynamic variables revealed a pattern strongly reminiscent of autoregulation (Fig 5). During the first two minutes of the task there was a marked increase in cardiac output while total peripheral resistance decreased. This pattern was partially reversed in the following three minutes; as anticipated, total peripheral resistance rose during the second phase of the experiment, while cardiac output partially returned to baseline. One might speculate that the decrease in cardiac output is secondary to a simultaneous increase in afterload, but this is unlikely as will be shown later. Similar autoregulatory events have been described by others for forearm muscle blood flow, where the time course of responses to continuous epinephrine infusion resembles the pattern of responses to mental stress (Barcroft et al 1960). Epinephrine itself usually increases during mental stress (Bonelli 1982). This would suggest that the increase in resistance during the second phase of task performance is indeed a nonspecific secondary feedback mechanism.

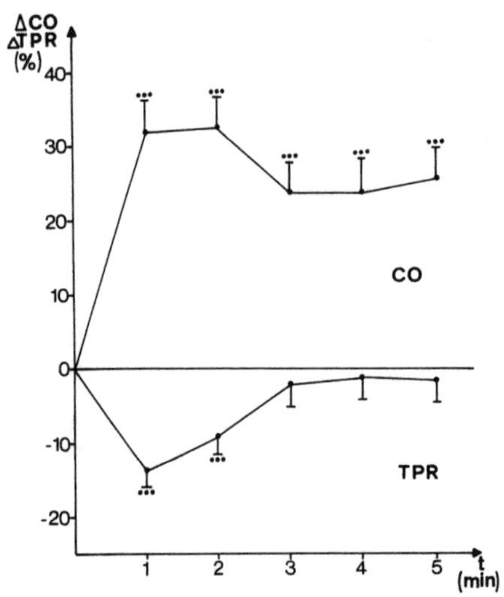

Fig 5. Average reactions of cardiac output (CO) and total peripheral resistance (TPR) to a modified mental arithmetic test in 28 young male healthy subjects. Reactivity is expressed as percent of baseline (mean ± SEM). Points denote significant changes from baseline (... $p<0.0\overline{0}1$).

Similar results were found in a larger sample of 59 men using our standard arithmetic stressor, although the amplitude of the vasodilatation and cardiac output increase was smaller (Fig 6, left panel). Unexpectedly, this pattern was not found in 67 female subjects (Fig 6, right panel). The time course of cardiac output was similar to that of men as was the initial vasodilatation. However, in contrast to men, women maintained the vasodilatation and diminution of total peripheral resistance. This argues against the notion that the decrease in cardiac output during the stress period is influenced by changes in total peripheral resistance. Another difference between the sexes was found in stroke volume. In both groups there was an initial decrease in stroke volume, corresponding to the initial peak in heart rate. Following this, women exhibited a continuous rise in stroke volume above the baseline, while in men stroke volume returned to basal levels. Similar differences were found when we compared women who never used oral contraceptives (n=24) with women on contraceptive drugs (n=23; Fig 7); women on contraceptives exhibited stronger vasodilatation and a greater increase of stroke volume. However, in neither group of women were signs of autoregulation observed.

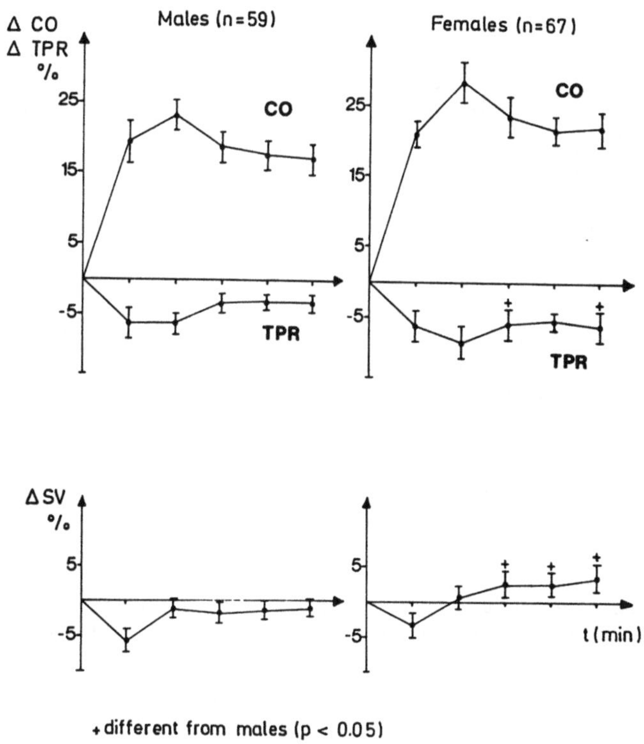

Fig 6. Average reactions of cardiac output (Δ CO), total peripheral resistance (Δ TPR) and stroke volume (Δ SV) to a mental arithmetic task, expressed in percent of baseline in male and female young healthy subjects (mean ± SEM). + denotes significant differences between the sexes ($p<0.05$).

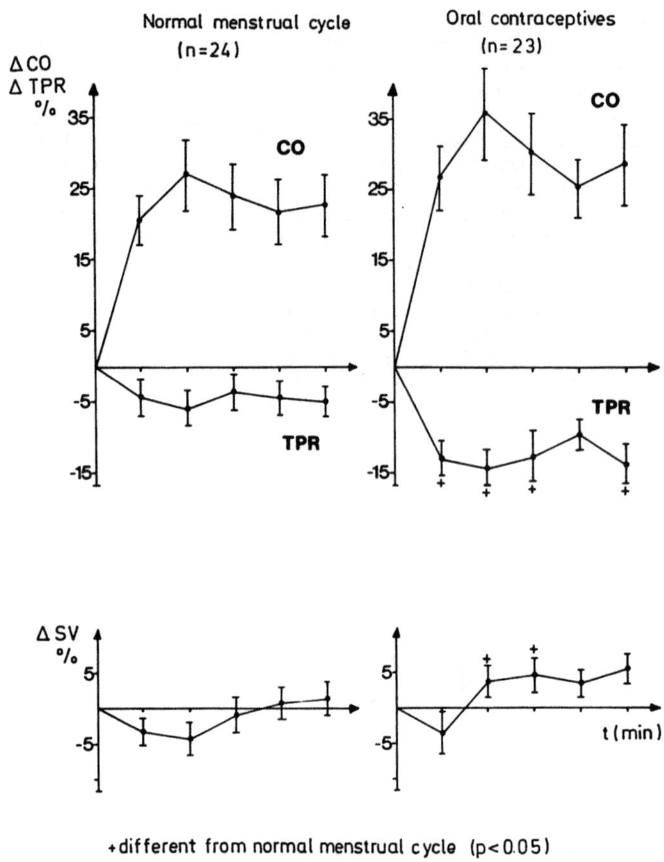

Fig 7. Average reactions of cardiac output (Δ CO), total peripheral resistance (Δ TPR) and stroke volume (Δ SV) to a mental arithmetic task, expressed as percent of baseline, in two groups of young female subjects (mean ± SEM). The subjects in the left panel had never used oral contraceptives, whereas the subjects on the right panel were on contraceptive drugs at the time of the stress test. + denotes group differences ($p<0.05$).

We conclude from these results that female sex hormones, in particular oestrogen, interfere with the autoregulatory response usually seen in men.

In summary, the following conclusions about the time course of cardiovascular reactions to mental stress can be drawn:

1. During mental arithmetic, a steady state is usually reached after two to three minutes of stress exposure. This justifies the use of averages across several minutes for calculating reactivity scores.

2. During shock avoidance and reaction time tasks, an initial peak

in systolic blood pressure, similar to that in heart rate, is often seen. In contrast, systolic and diastolic blood pressure tend to increase over the first minute of exposure to mental arithmetic.

3. The physiological sequence of events must be taken into account when evaluating and interpreting the time course of the stress response. When changes in cardiovascular function under stress are reported, they should be related to the time scale of measurements.

4. The magnitude of the initial peak in heart rate seems to be indicative of sympathetic drive to the heart and functional hyperreactivity.

5. Analysis of the time course of responses may help to decipher the control mechanisms relevant to the stress reaction (baroreflex, autoregulation, oestrogen).

THE INFLUENCE OF GENDER

It was noted in an early study by von Eiff (1970) that sex differences exist in cardiovascular reactivity to mental stress. Blood pressure reactions were smaller in women, independently of baseline differences. Similarly, Frankenhaeuser et al (1978) reported that women exhibited lower epinephrine responses during a six-hour examination than men. However, there are data which indicate that these differences may be influenced, at least in part, by differences in socioeconomic status between men and women. Sex differences in reactivity may be less obvious when women live in a 'male' environment (Neus et al 1985; Collins and Frankenhaeuser 1978).

This point can be illustrated by a recent investigation in which we compared male and female subjects screened from a city population, and female students (Neus et al 1985; Fig 8). In the city population, it was found that females exhibited smaller increases in blood pressure than males. The group of female students reacted as strongly as the male city population, suggesting that socioeconomic factors may indeed mask hormonal effects. However, the proportion of women taking oral contraceptives was much lower in the group of students (15%) than in the female city population (50%); furthermore, in the latter group, women taking oral contraceptives showed smaller systolic blood pressure reactions than women who did not ($p < 0.05$). Therefore, the differences between the female groups could also be attributed to the difference in contraceptive practice.

These data are consistent with the hypothesis that oestrogen, whether released endogenously or introduced artificially, attenuates cardiovascular reactions to mental stress. Besides the differences between both sexes and the influence of oral contraceptives, this has also been demonstrated by analysing the effect of menstrual cycle, pregnancy, and introduction of oestrogen in women after menopause (von Eiff et al 1971; von Eiff and Piekarski 1977). Irrespective of what causes these effects in terms of physiological agents, it is clear that the role of gender and sex hormones must be taken into account when women participate in psychophysiological studies.

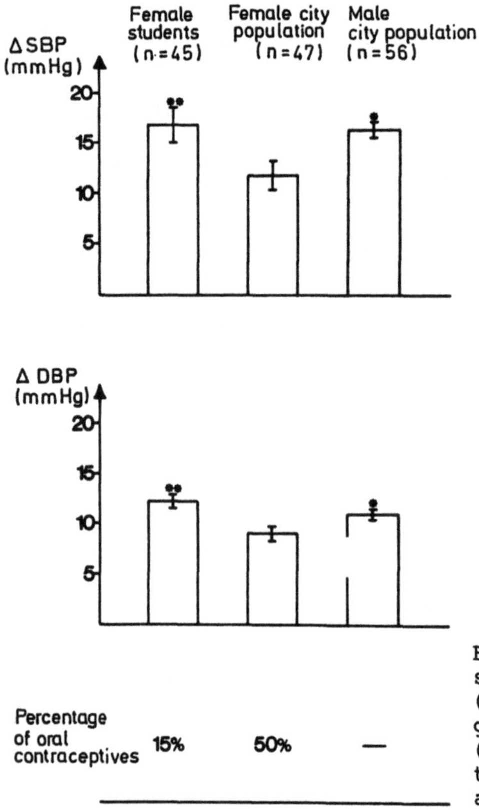

Fig 8. Average reactions of systolic (Δ SBP) and diastolic (Δ DBP) blood pressure in three groups of healthy young subjects (mean ± SEM). Differences from the 'female city population' group are tested (** $p<0.01$, * $p<0.05$).

ADAPTATION TO REPEATED TESTING

Repeated stress testing is necessary when prospective or intervention studies are performed. In such instances, an ideal stress test would exhibit no systematic adaptation (ie, no change in mean reactivity scores), and a high test-retest correlation. The diagnosis of hyperreactivity by use of invariable cutpoints would be very advantageous. But although test-retest correlations are usually in the medium to upper range (see Chapter 8), this requirement is usually not fulfilled by conventional stress test procedures.

We first became aware of the problem of adaptation when we performed a therapeutic trial on the effects of long-term ß-blockade on the stress response in hypertensives. We noticed that even during the placebo phase, cardiovascular reactions were so greatly attenuated that the effect of ß-blockers could not be properly analysed (Friedrich et al 1981). The adaptation effect was stronger than was predicted from earlier investigations of normotensives. A methodological study was therefore performed on 12 normotensives and 12 hypertensives, with repeated stress testings within one week (Neus et al 1981). The expected differences between normotensives and hypertensives in blood pressure reactivity emerged in the first test, but these effects disappeared when the subjects were retested. The pattern was due to marked adaptation in the group of hypertensives (ie, the high reactors). The same was true for heart rate reactivity. One conclusion to be drawn from this study was that hyperreactivity in hypertensives cannot be attributed exclusively to structural changes in the vasculature, because these changes are obviously the same in repeated tests. On the other hand, this result clearly points up a methodological problem, in that the phenomenon of hyperreactivity disappears on retesting.

Adaptation phenomena are not limited to studies of hypertensives. In a recent experiment, retesting was carried out in 60 boys after a year's interval. Heart rate reactivity was more reliable than blood pressure reactivity; therefore we divided the sample into two groups by median split on heart rate reactivity. As can be seen from Fig 9, a clear adaptation phenomenon again occurred in the high reactivity group. This was not the case for the control group of normal reactors. Nevertheless, group differences were

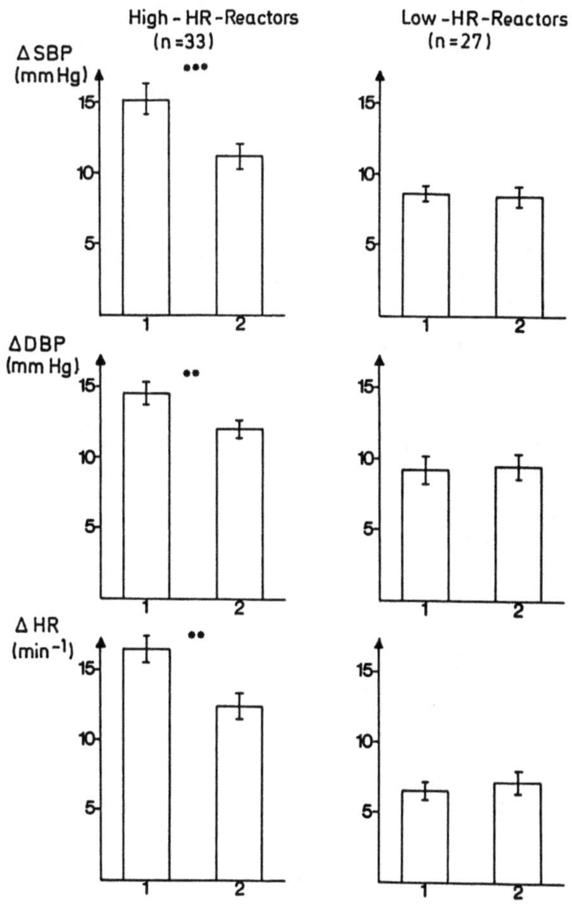

Fig 9. Average cardiovascular reactions in 60 boys (10-12 years of age in test 1) tested two times at a year's interval. The sample was divided into two groups by median split on HR reactions to the first test (mean$\pm$SEM). Differences between the reactions to test 1 and 2 were analysed (*** $p < 0.001$, ** $p < 0.01$).

maintained. Similar results have been reported in studies by Manuck et al (Manuck and Garland 1980; Giordani et al 1981), where adaptation was especially prominent in the high reactor groups. It should be noted, however, that - as in our results - group differences were maintained.

One possibility is that these patterns simply reflect 'regression to the mean'. It might be argued that the observations are statistical artefacts rather than physiologically relevant. If the hyperreactivity diagnosed in the first test is an error (resulting from measurement errors or situational factors), a weaker reaction to the second test can be expected. From this point of view, 'adaptation' phenomena are to be expected, particularly in hyperreactive subjects. However, the results reported for hypertensives cannot be explained in this way, because it is extremely unlikely that their initial hyperreactivity is an error. In our experiments on children we defined the groups on the basis of heart rate reactivity; therefore the adaptation in blood pressure observed in the high heart rate reactor group cannot be attributed to measurement errors. Nevertheless, it is possible that situational factors also contribute to the pattern of results.

In conclusion, adaptation to repeated testing, which occurs most markedly in highly reactive subjects, imposes a severe problem in any long-term study. An experimental stress test which did not show adaptation effect would be highly desirable. On the other hand, test-retest correlations are in the medium to upper range, indicating that relative reactivity is a stable characteristic.

REFERENCES

Barcroft H, Brod J, Hejl Z, Hirsjärvi EA, Kitchin AH (1960) The mechanisms of the vasodilatation in the forearm muscle during stress (mental arithmetic). Clin Sci 19: 577-586

Bonelli J (1982) Stress, catecholamines and beta-blockade. Acta Med Scand, Suppl 660: 214-218

Bristow JD, Honour AJ, Pickering GW, Sleight P, Smyth HS (1969) Diminished baroreflex sensitivity in high blood pressure. Circulation Res 24: 48-54

Brod J, Fencl V, Hejl Z, Jirka J (1959) Circulatory changes underlying blood pressure elevation during acute emotional stress (mental arithmetic) in normotensive and hypertensive subjects. Clin Sci 18: 269-279

Brod J, Cachovan M, Bahlmann J, Bauer GE, Celsen B, Sippel R, Hudeshagen H, Feldmann U, Rienhoff O (1979) Haemodynamic changes during acute emotional stress in man with special reference to the capacitance vessels. Klin Wochenschr 57: 555-565

Collins A, Frankenhaeuser M (1978) Stress responses in male and female engineering students. J Hum Stress 4: 43-48

Conway J, Boon N, Jones JV, Sleight P (1983) Involvement of the baroreceptor reflexes in the changes in blood pressure with sleep and mental arousal. Hypertension 5: 746-748

De Good A (1975) Cognitive control factors in vascular stress responses. Psychophysiology 12: 399-401

von Eiff AW (1970) The role of the autonomic nervous system in the etiology and pathogenesis of essential hypertension. Jpn Circ J 34: 147-153 von Eiff AW, Czernik A, Zanders H (1969) Zur medikamentösen Beeinflussung der Sympathikus-hyperaktivität. Parasympathomimeticum und Beta-Receptorenblocker im kurzdauernden pharmakologischen Experiment am Menschen. Klin Wochenschr 47: 701-708

von Eiff AW, Plotz EJ, Beck KJ, Czernik A (1971) The effect of estrogens and progestins on blood pressure regulation of normotensive women. Amer J Obs & Gynec 109: 887-892

von Eiff AW, Piekarski C (1977) Stress reactions of normotensives and hypertensives and the influence of female sex hormones on blood pressure regulation. In: de Jong W, Provoost AP, Shapiro AP (eds) Hypertension and Brain Mechanisms (Progress in Brain Research, Vol 47). Elsevier, Amsterdam, p289

Falkner B, Onesti G, Angelakos ET, Fernandes M, Langman, C (1979) Cardiovascular response to mental stress in normal male adolescents with hypertensive parents. Hypertension 1: 23-30

Forsman L, Lindblad, LE (1983) Effects of mental stress on baroreceptor-mediated changes in blood pressure and heart rate and on plasma catecholamines and subjective responses in healthy men and women. Psychosom Med 45: 435-445

Frankenhaeuser M (1983) The sympathetic-adrenal and pituitary-adrenal response to challenge: comparison between the sexes. In: Dembroski TM, Schmidt TH and Blümchen G (eds) Biobehavioral Bases of Coronary Heart Disease. Karger, Basel, p91

Frankenhaeuser M, Rauste-von Wright M, Collins A, von Wright J, Sedvall G, Swahn CG (1978) Sex differences in psychoneuroendocrine reactions to examination stress. Psychosom Med 40: 334-343

Friedrich G, Langewitz W, Neus H, Schirmer G, Thönes M (1981) Der emotionale Belastungstest in der klinisch-therapeutischen Prüfung von Antihypertensiva. Verh dtsch Ges inn Med 87: 551-554

Giordani B, Manuck SB, Farmer JF (1981) Stability of behaviorally induced heart rate change in children after one week. Child Devel 52: 533-532

Guyton AC (1980) Circulatory Physiology. III: Arterial pressure and hypertension. Saunders, Philadelphia

Guyton AC, Jones CE, Coleman TG (1973) Circulatory Physiology. I: Cardiac output and its regulation. Saunders, Philadelphia

Korner PI, Shaw J, West MJ, Uther JB (1974) 'Steady state' properties of the baroreceptor-heart rate reflex in essential hypertension in man. Clin Exper Pharmacol Physiol 1: 65-73

Kubicek WG, Kottke FJ, Ramos MU, Patterson RO, Witsoe DA, Labree JW, Remole W, Layman TE, Schoening H, Garamele JT (1974) The Minnesota impedance cardiograph. Theory and applications. Biomed Eng 9: 410-416

Mancia G, Ferrari A, Gregorini L, Ludbrook J, Zanchetti A (1978) Baroreceptor control of heart rate in man. In: Schwartz PJ, Brown AM, Malliani A, Zanchetti A (eds) Neural Mechanisms in Cardiac Arrhythmias. Raven Press, New York, p323

Manuck SB, Garland FN (1980) Stability of individual differences in cardiovascular reactivity: a 13-month follow-up. Physiol Behav 24: 621-624

Melville DI, Raftery EB (1981) Blood pressure changes during acute mental stress in hypertensive subjects using the Oxford intraarterial system. J Psychosom Res 25: 487-497

Neus H, von Eiff AW, Friedrich G, Heusch G, Schulte W (1981) Das Problem der Adaptation in der klinisch-therapeutischen Hypertonieforschung. Dtsch Med Wschr 196: 622-624

Neus H, von Eiff AW, Friedrich G, Rüddel H, Schulte W (1983) Der Einfluss von Betarezeptorenblockern und Diuretika auf die Reaktivitt des Blutdrucks. Z Kardiol 72: 617-621
Neus H, von Eiff AW (in press) Lärm und Hypertonie. In: Rosenthal J, (ed) Arterielle Hypertonie (2nd ed). Springer, Berlin, Heidelberg, New York
Neus H, Otten H, Rüddel H, Schmieder R, Schulte W and von Eiff AW (1985) Cardiovascular and emotional reactions to mental stress. In: Orlebeke JF, Mulder G, van Doornen LJP (eds) Psychophysiology of Cardiovascular Control. Plenum Press, New York
Nyberg G (1976) Blood pressure and heart rate response to isometric exercise and mental arithmetic in normotensive and hypertensive subjects. Clin Sci 51: 681s-685s
Obrist PA, Gaebelein CJ, Teller ES, Langer AW, Grignolo A, Light KC, McCubbin JA (1978) The relationship among heart rate, carotid dp/dt and blood pressure in humans as a function of the type of stress. Psychophysiology 15: 102-115
Rüddel H, Langewitz W, McKinney, ME, Todd GL, Buell JC, Eliot RS (in press) Haemodynamic responses during the Type A interview: a comparison with mental challenge and a clinical interview. J Autonom Nerv System
Schirmer G (1981) Reaktionen der Herzfrequenz und ihrer Variabilität im Zusammenhang mit der vegetativen Innervation des Herzens. PhD dissertation, University of Bonn
Schulte W, Neus H (1979) Bedeutung von Streßreaktionen in der Hypertoniediagnostik. Herz/Kreislauf 11: 541-546
Schulte W, Neus H, von Eiff AW (1981) Blutdruckreaktivität bei unkomplizierten Formen des Hochdrucks. Klin Wochenschr 59: 1243-1249
Schulte W, Neus H (1983) Haemodynamics during emotional stress in borderline and mild hypertension. Europ Heart J 4: 803-809

CHAPTER 8

Adaptation Effects and the Stability of Physiological Responses to Repeated Testing

Michael Myrtek

Forschungsgruppe Psychophysiologie, Universität Freiburg, Belfortstrasse 20, D-7800 FREIBURG i.Br., Federal Republic of Germany

SUMMARY

Adaptation effects and stability coefficients of responses during repeated psychophysiological testing were investigated in a sample of 48 male students. In the repeated examinations at one week intervals, the following stress tests were used: Reaction Time Test with contingent performance feedback, Cold Pressor Test, and a Number Sequence Test with monetary incentives. Heart rate, stroke volume, cardiac output, ventricular ejection time, pulse wave velocity, blood pressure, respiration rate, tidal volume, respiratory minute volume, oxygen (O_2) uptake, O_2 equivalent and respiratory quotient were measured. Reactivity of subjects was measured using a number of different change scores: arithmetical differences between stress test and resting levels, distances from the first principal axis (AHA), and autonomic lability scores (ALS).

Significant adaptation effects were found in a number of analyses, depending on the physiological parameter studied and to a lesser extent on the nature of the stress test evaluated. Change scores showed only modest adaptation effects, varying with the different stress tests. The magnitude of stability coefficients again depended on the physiological parameter in question. Change scores showed some remarkably high coefficients. No substantial differences were evident in the magnitude of stability coefficients emerging with the different change scores analysed.

In a second study with 54 male cardiac patients, correlations between single stress tests and habitual responses (defined as mean scores over 17 repeated test sessions) were investigated. The results of this study show that values derived from the second measurement session are more reliable predictors of habitual responses than those produced in the first session. A second examination is therefore recommended, since this may overcome adaptation effects and enhance diagnostic validity.

INTRODUCTION

Experience has shown that the results of physiological assessments taken during the first medical examination should be evaluated with caution. A patient initially diagnosed as hypertensive may well show normal blood pressure values in a second or third examination. Factors such as the medical apparatus used and the unfamiliarity of the laboratory setting may contribute to variations in physiological measures. These factors may also influence the results of repeated stress testing. It is methodologically important that the nature and intensity of these disturbing factors is known. The present paper addresses the following questions:

(1) What physiological parameters and which types of stress tests are most strongly influenced by adaptation effects?

(2) Are there differences in adaptation effects between rest, stress test (strain), and change (reactivity) scores?

(3) How high are the stability coefficients of different scores for different stress tests? Stability coefficients are defined as the product-moment correlations between scores collected on different occasions within a group of subjects.

(4) What is the correlation between different types of change score?

(5) Can inferences be made about an individual's habitual pattern of physiological resonse from the results of an initial examination?

STUDY 1: ADAPTATION EFFECTS AND TEST-RETEST STABILITY

As part of an experiment designed to examine psychophysiological activation, 48 male student volunteers (mean age 25.9 years, SD = 3.6) were examined twice at an interval of one week. All strain values were related to a baseline measurement (rest phase) of two minutes' duration, that occurred between the Cold Pressor Test and the Number Sequence Test. This rest value was chosen as the baseline instead of the initial baseline, because the latter showed somewhat lower stability coefficients in many variables (difference of mean stability coefficients = .04).

The following stress tests were used:

Reaction Time Test (RT). The duration of recording was 120 sec. In this test the reaction time exhibited by the subject was compared to his mean score in a pretest. When the subject responded faster than this reaction time threshold, he was given contingent acoustic feedback, thus increasing motivation to perform well.

Cold Pressor Test (CP). The duration of recording was 60 sec. Subjects were requested to immerse their right hand in cold water (4°C) for one minute.

Number Sequence Test (NS). The duration of recording was 120 sec. In this test, various series were selected from the Intelligence Structure Test (IST) designed by Amthauer (1955). Subjects were promised and later paid a bonus in addition to their participation fee that corresponded to the number of properly solved series.

The following physiological variables were recorded:

(1) Heart rate (bpm): conventional ECG based on impedance cardiography (IFM impedance cardiograph, Model 400).

(2) Stroke volume (ml): based also on impedance cardiography.

(3) Cardiac output (litre/min): based on impedance cardiography.

(4) Left ventricular ejection time (LVET in msec): based on impedance cardiography.

(5) Pulse wave velocity (cm/sec): here, pulse transit time was measured between the R wave of the ECG and a point 20% up the ascending face of the femoralis pulse curve (pulse transducer from the Boucke-Brecht Company). Arterial length was determined by measuring the distance between jugulum-navel-femoralis (mid point of the transducer).

(6) Blood pressure (mmHg): systolic blood pressure was determined at the first Korotkoff sound, diastolic blood pressure from the point of distinct muffling using a mercury manometer.

(7) Respiratory rate (cycles/min): measured using the pneumotest system (Jaeger Inc, Würzburg).

(8) Tidal volume (ml): also based on the pneumotest system.

(9) Respiratory minute volume (litres/min): by the pneumotest system.

(10) O_2 uptake (ml/min): gas analysis using an Oxytest O_2-analyser (Jaeger Inc).

(11) O_2 equivalent: respiratory minute volume (ml/min)/O_2 uptake (ml/min).

(12) Respiratory quotient: carbon dioxide output (ml/min)/ O_2 uptake (ml/min). Analysis of CO_2 was conducted using a Uras CO_2-analyser (Hartmann & Braun Co, Frankfurt).

TYPES OF CHANGE SCORES

In addition to the simple arithmetic differences (D) between strain (final) values and rest (initial) values, we used the autonomic lability score (ALS), first devised by Lacey and Lacey (1962). The distance from the first principal axis, referred to as AHA, was also calculated; this controls for the dependency of change scores on initial values (see Myrtek et al, 1977; Fahrenberg et al, 1983). Since this latter score is not generally familiar, a brief explanation may be valuable.

The ALS is widely used in the literature to denote change, because it formally eliminates initial-value dependency. Unfortunately, the ALS does not take into account the important a (a-b) effect. The notion of an a (a-b) effect implies that for any correlation between two values that share a common term, eg, for a and (a-b) or for x and (x-y), the correlation coefficient cannot be equal to zero, if a and b are independent. It follows that whenever the correlation coefficient $r_{xy} < 1.00$, any initial value dependency is contaminated by the negative correlation r_{dx} (where d = y-x), resulting from the a (a-b) effect. The following variance components contribute to the a (a-b) effect. Firstly, variance due to differences in reactivity exhibited by subjects who have the same initial value. This means that different subjects respond differently to the same stimulus. Secondly, variance due to the 'noise' of measurement (ie, methodological error; see Myrtek, 1984). Since the a (a-b) effect is a statistical artefact, its influence on any real negative correlation between initial values and differences (r_{dx}) must be controlled. A method of eliminating the dependency of change scores on initial values due to the a (a-b) effect involves calculating the distance from the first principal axis, referred to as AHA. Based on the combined data points of initial x_i and final y_i values for all subjects on variable j, a slope is fitted by the least squares method:

$$AHA_{ij} = (y_{ij} - ß_j * x_{ij})/(1 + ß^2_j)^{1/2}$$

where $ß_j$ is the slope of the first principal component.

ADAPTATION EFFECTS IN REST AND STRAIN VALUES

Table 1 presents the adaptation effects for the cardiovascular variables and Table 2 shows the results for the respiratory variables. It can be seen that adaptation effects in rest and strain values vary according to the parameter studied, and to a lesser degree with the kind of stress test used. Heart rate, systolic blood pressure and O_2 equivalent are especially influenced by adaptation effects; in the second examination, subjects showed significantly lower values of heart rate and systolic blood pressure under most conditions, as well as a higher O_2 equivalent. The Number Sequence Test and Reaction Time Test were more susceptible to adaptation effects than the Cold Pressor Test or rest phase. In contrast, little adaptation was observed for diastolic blood pressure indices of myocardial performance or other respiratory parameters.

The stability coefficients also vary with the physiological variables employed and the type of stress test used. The individual stability coefficients were transformed using Fisher's z, so that the mean stability coefficient could be calculated for each parameter across experimental conditions. The results of this analysis are summarised in Table 3. Relatively high stability coefficients are evident for stroke volume, heart rate, systolic BP and cardiac output. Lower stability coefficients were found for the respiratory variables, especially for O_2 equivalent and respiratory quotient.

ADAPTATION EFFECTS AND STABILITY COEFFICIENTS FOR DIFFERENCE SCORES

Tables 4 and 5 present the results for arithmetic difference scores. Compared with rest and strain values, adaptation effects are less pronounced with arithmetic difference scores, and depend on the type of stress test used. Only the Number Sequence Test showed significant adaptation effects consistently (for heart rate, stroke volume, left ventricular ejection time, pulse wave velocity and systolic blood pressure). Values recorded during the Cold Pressor Test and Reaction Time Test did not vary across occasions.

The stability coefficients of difference scores vary with the parameter studied and the kind of stress test used. Some

Table 1. Adaptation effects and stability coefficients of cardiovascular rest and strain scores to repeated testing.

Physiological variable	Exp. Cond.	Means T1	T2	Diff. T1-T2	p	Stab.
Heart rate	Rest	68.8	66.8	2.0	.004	.62***
	RT	77.4	73.7	3.7	.004	.68***
	CP	78.8	77.9	0.9	.227	.79***
	NS	90.0	82.8	7.2	.000	.78***
Stroke volume	Rest	118.7	119.4	-0.7	.598	.74***
	RT	114.8	116.1	-1.3	.611	.72***
	CP	111.9	109.3	2.6	.611	.76***
	NS	102.2	108.1	-5.9	.030	.72***
Cardiac output	Rest	8.1	7.9	0.2	.260	.72***
	RT	8.8	8.5	0.3	.217	.64***
	CP	8.6	8.3	0.3	.257	.66***
	NS	9.1	8.9	0.2	.453	.70***
Left ventricular ejection time	Rest	303.1	306.2	-3.1	.062	.51***
	RT	289.3	296.8	-7.5	.006	.58***
	CP	292.6	295.3	-2.7	.202	.59***
	NS	281.7	291.5	-9.8	.001	.78***
Pulse wave velocity	Rest	284.8	280.6	4.2	.053	.67***
	RT	306.7	295.7	11.0	.009	.67***
	CP	304.7	298.6	6.1	.101	.63***
	NS	327.2	312.4	14.8	.003	.64***
BP systolic	Rest	122.5	118.2	4.3	.000	.73***
	RT	128.9	123.0	5.9	.000	.68***
	CP	142.4	139.1	3.3	.038	.67***
	NS	135.6	127.9	7.7	.000	.69***
BP diastolic	Rest	77.8	76.9	0.9	.446	.48***
	RT	76.8	77.1	-0.3	.817	.62***
	CP	91.5	93.1	-1.6	.275	.62***
	NS	82.9	80.8	2.1	.046	.60***

Measurements of T_1 and T_2 with an interval of 1 week (N = 48). Experimental conditions: Rest, RT = Reaction Time Test, CP = Cold Pressor Test, NS = Number Sequence Test, p = significant levels of a t-test for the mean difference $T_1 - T_2$, Stab. = stability coefficients. Level of significance: *$p \leq .05$, ** $p \leq .01$, *** $p \leq .001$.

Table 2. Adaptation effects and stability coefficients of respiratory rest and strain scores for repeated testing.

Physiological variable	Exp. Cond.	Means T1	T2	Diff. T1-T2	p	Stab.
Respiratory rate	Rest	12.0	12.3	-0.3	.376	.72***
	RT	16.4	17.1	-0.7	.170	.60***
	CP	15.8	15.4	0.4	.583	.56***
	NS	17.8	17.3	0.5	.443	.31*
Tidal volume	Rest	527.3	519.6	7.7	.712	.53***
	RT	467.6	477.6	-10.0	.510	.56***
	CP	510.6	550.5	-39.9	.077	.52***
	NS	485.3	490.6	-5.3	.743	.59***
Respiratory minute volume	Rest	6.0	6.2	-0.2	.459	.59***
	RT	7.6	8.1	-0.5	.074	.59***
	CP	7.9	8.3	-0.4	.258	.49***
	NS	8.4	8.3	0.1	.781	.52***
O_2 uptake	Rest	206.0	202.4	3.6	.600	.59***
	RT	239.6	247.2	-7.6	.361	.52***
	CP	251.8	252.9	-1.1	.887	.51***
	NS	271.3	261.9	9.4	.280	.53***
O_2 equivalent	Rest	29.2	30.4	-1.2	.012	.30*
	RT	31.5	32.6	-1.1	.027	.37**
	CP	31.7	33.1	-1.4	.018	.31*
	NS	31.0	31.8	-0.8	.077	.47***
Respiratory quotient	Rest	0.73	0.74	-0.01	.602	.43**
	RT	0.80	0.81	-0.01	.747	.45**
	CP	0.79	0.80	-0.01	.268	.35*
	NS	0.77	0.77	0.00	.702	.37**

Further explanation see Table 1.

Table 3. Mean stability coefficients of rest, strain and change scores for cardiovascular and respiratory variables.

Physiological variable	Rest scores	Strain scores	Change scores D	Change scores AHA	Change scores ALS
Heart rate	.62	.75	.82	.75	.83
Stroke volume	.74	.73	.63	.62	.60
Cardiac output	.72	.67	.57	.62	.59
LVET	.51	.66	.76	.64	.75
Pulse wave velocity	.67	.65	.65	.61	.61
BP systolic	.73	.68	.46	.46	.47
BP diastolic	.48	.61	.12	.07	.25
Respiratory rate	.72	.50	.53	.53	.46
Tidal volume	.53	.56	.57	.57	.57
Respiratory minute volume	.59	.53	.52	.54	.53
O_2 uptake	.59	.52	.48	.52	.49
O_2 equivalent	.30	.39	.28	.27	.33
Respiratory quotient	.43	.39	.39	.40	.39
Mean	.61	.60	.55	.53	.55

Values are based on the single stability coefficients of Tables 1, 2, 5 and 6, computing the mean coefficient (following z transformation) for the three experimental stress tests.

Table 4. Adaptation effects and stability coefficients of cardiovascular change scores (differences between strain and rest scores) for repeated testing.

Physiological variable	Exp. Cond.	Means T1	Means T2	Diff. T1-T2	p	Stab.
Heart rate	RT	8.3	6.9	1.4	.051	.69***
	CP	10.0	11.1	-1.1	.091	.90***
	NS	20.8	16.0	4.8	.000	.83***
Stroke volume	RT	-3.1	-3.3	0.2	.803	.55***
	CP	-6.8	-10.1	3.3	.105	.68***
	NS	-15.8	-11.3	-4.5	.019	.64***
Cardiac output	RT	0.7	0.6	0.1	.410	.58***
	CP	0.5	0.4	0.1	.610	.54***
	NS	1.0	1.0	0.0	.724	.60***
Left ventricular ejection time	RT	-13.0	-9.4	-3.6	.054	.35*
	CP	-10.4	-10.9	0.5	.794	.80***
	NS	-20.6	-14.7	-5.9	.002	.91***
Pulse wave velocity	RT	22.5	15.1	7.4	.001	.61***
	CP	19.9	18.0	1.9	.488	.71***
	NS	44.3	31.9	12.4	.000	.61***
BP systolic	RT	6.3	4.8	1.5	.154	.35*
	CP	19.8	20.9	-1.1	.506	.61***
	NS	13.1	9.7	3.4	.016	.40**
BP diastolic	RT	-1.0	0.2	-0.8	.301	-.06
	CP	13.7	16.5	-2.8	.138	.36*
	NS	5.1	3.9	1.2	.332	.04

Further explanation see Table 1.

Table 5. Adaptation effects and stability coefficients of respiratory change scores (differences between strain and rest scores) for repeated testing.

Physiological variable	Exp. Cond.	Means T1	Means T2	Diff. T1-T2	p	Stab.
Respiratory rate	RT	4.4	4.7	-0.3	.503	.56***
	CP	3.8	3.2	0.6	.176	.55***
	NS	5.8	4.9	0.9	.146	.49***
Tidal volume	RT	-59.7	-42.0	-17.7	.317	.64***
	CP	-16.7	27.1	-43.8	.051	.48***
	NS	-42.0	-28.9	-13.1	.404	.59***
Respiratory minute volume	RT	1.6	1.9	-0.3	.210	.49***
	CP	1.9	2.2	-0.3	.511	.44**
	NS	2.4	2.2	0.2	.390	.61***
O_2 uptake	RT	33.6	44.8	-11.2	.192	.38**
	CP	45.8	51.0	-5.2	.639	.47***
	NS	65.4	60.3	5.1	.541	.58***
O_2 equivalent	RT	2.3	2.2	0.1	.703	.24
	CP	2.5	2.6	-0.1	.973	.17
	NS	1.8	1.3	0.5	.093	.43**
Respiratory quotient	RT	0.07	0.07	0.00	.893	.52***
	CP	0.05	0.06	-0.01	.370	.24
	NS	0.04	0.03	0.01	.283	.40**

Further explanation see Table 1.

coefficients are remarkably high, eg, left ventricular ejection time during the Number Sequence Test (.91) and heart rate during the Cold Pressor Test (.90). Indeed, these coefficients are higher than the corresponding stability coefficients for rest or strain scores. Table 3 summarises the mean stability coefficients for difference scores for each parameter. Compared with the rest or strain analysis, coefficients for difference scores for heart rate and ventricular ejection time are higher. The coefficients for blood pressure are considerably lower than the corresponding rest or strain scores, while other parameters show similar stability in the two analyses.

COMPARISON OF DIFFERENT CHANGE SCORES

Tables 6 and 7 present comparisons between the three forms of change score explored in this study. Intercorrelations between difference scores (D), distances from the first principal axis (AHA), and autonomic lability scores (ALS) are presented, together with stability coefficients for the three change scores seperately. The correlations between D and ALS are uniformly high for the cardiovascular parameters, ranging from .84 to 1.00. The correlations between D and AHA scores and between AHA and ALS scores are more variable. The respiratory parameters show high correlations between D and AHA scores, somewhat lower correlations between D and ALS, and low correlations between AHA and ALS scores. Turning to the stability coefficients, D and ALS scores show a similar pattern while AHA scores deviate somewhat. Overall, however, there are no striking differences for the various change scores on the stability coefficients (see Table 3).

STUDY 2: THE COMPARISON OF SINGLE MEASUREMENTS WITH HABITUAL RESPONSES

A second study was conducted in order to examine the relationship between responses during initial stress testing and the 'habitual' values collected with repeated testing. A sample of 36 patients suffering from myocardial infarction (mean age 43.6 years) and 18 patients with functional cardiovascular disorders (mean age 43.8 years) was collected. A longitudinal study was conducted with the patients during a rehabilitation treatment that lasted six weeks (see Myrtek, 1984). A total of 17 examinations were carried out. The experimental protocol consisted of a rest phase lasting 120 sec, and maximal hyperventilation for a 30 sec period. The physiological parameters recorded and apparatus used were the same as in the student sample detailed earlier.

Habitual scores were determined by calculating mean responses over all test sessions. The scores recorded during individual test sessions were then correlated with habitual scores in order to obtain an estimate of the predictive validity of single measurements. The first, second and ninth test sessions have been selected for this demonstration. The results are presented in Table 8. This shows the correlations between the first, second or ninth session on the one hand, and the habitual scores on the other. It is evident that systolic blood pressure can be predicted adequately from the first session responses, since subsequent measures do not increase the correlation with habitual scores significantly. This finding does not, however, hold for diastolic blood pressure. In the case of diastolic blood pressure, the prediction of the habitual scores is substantially enhanced by using values from the second session. Generally across the other physiological parameters shown in Table 8, prediction is increased when responses from the second session are employed.

According to classical test theory, the test-retest correlation coefficient may be viewed as an estimate of the common variance between the test and its repetition. Averaging over all variables in this sample, the common variance between responses to the first examination and habitual scores is 61%. Common variance increases to 78% for the second session, and to 80% for responses from the ninth session.

Table 6. Correlations between different cardiovascular change scores (differences, D, distances from the first principal axis, AHA, and autonomic lability scores, ALS) and stability coefficients for each type of change score.

		Correlations			Stability coefficients		
Physiological variable	Exp. cond.	D-AHA	D-ALS	AHA-ALS	D	AHA	ALS
Heart rate	RT	.75	.98	.87	.69	.75	.72
	CP	.66	.99	.76	.90	.77	.90
	NS	.61	1.00	.55	.83	.74	.82
Stroke volume	RT	1.00	.98	.97	.55	.55	.52
	CP	.95	1.00	.96	.68	.68	.70
	NS	.99	.88	.93	.64	.61	.55
Cardiac output	RT	.92	.99	.96	.58	.64	.60
	CP	.90	.98	.97	.54	.61	.59
	NS	.96	.98	.88	.60	.61	.58
Left ventricular ejection time	RT	.95	.99	.89	.35	.36	.35
	CP	.90	1.00	.87	.80	.77	.79
	NS	.64	1.00	.57	.91	.71	.90
Pulse wave velocity	RT	.63	.84	.95	.61	.46	.49
	CP	.82	.97	.93	.71	.77	.73
	NS	.68	.96	.86	.61	.53	.57
BP systolic	RT	.90	1.00	.93	.35	.34	.35
	CP	.87	.98	.75	.61	.63	.62
	NS	.83	1.00	.84	.40	.39	.40
BP diastolic	RT	1.00	.93	.92	-.06	-.06	.07
	CP	.81	.97	.63	.36	.21	.47
	NS	1.00	.93	.93	.04	.05	.18

Level of significance: .28 = $p \leq .05$, .37 = $p \leq .01$, .46 = $p \leq .001$. Further explanation see Table 1.

Table 7. Correlations of different respiratory change scores (differences, D, distances from first principal axis, AHA, and autonomic lability scores, ALS) and corresponding stability coefficients.

		Correlations			Stability coefficients		
Physiological variable	Exp. cond.	D-AHA	D-ALS	AHA-ALS	D	AHA	ALS
Respiratory rate	RT	.99	.91	.83	.56	.55	.54
	CP	.99	.93	.89	.55	.56	.51
	NS	1.00	.86	.82	.49	.49	.30
Tidal volume	RT	.94	.74	.93	.64	.64	.64
	CP	.99	.81	.89	.48	.47	.50
	NS	.93	.76	.94	.59	.58	.56
Respiratory minute volume	RT	.92	.95	.75	.49	.50	.54
	CP	.85	.91	.57	.44	.49	.47
	NS	.89	.80	.43	.61	.62	.58
O_2 uptake	RT	.99	.91	.84	.38	.41	.42
	CP	.89	.93	.65	.47	.52	.49
	NS	.98	.81	.70	.58	.62	.55
O_2 equivalent	RT	.99	.95	.92	.24	.22	.29
	CP	.98	.98	.93	.17	.16	.19
	NS	1.00	.90	.91	.43	.43	.50
Respiratory quotient	RT	.85	.98	.74	.52	.52	.51
	CP	.90	.99	.85	.24	.28	.24
	NS	.89	.98	.77	.40	.40	.39

Level of significance: .28 = $p \leq .05$, .37 = $p \leq .01$, .46 = $p \leq .001$. Further explanation see Table 1.

Table 8. Correlation coefficients between habitual Tt (means of 17 repeated examinations in a sample of N = 54 cardiac patients) and values of the first T1, second T2, or ninth T9 examination. Experimental conditions: rest and hyperventilation (HV) test.

Physiological Variable	Exp. Cond.	Correlations		
		T1-Tt	T2-Tt	T9-Tt
Heart rate	Rest	.50	.73	.81
	HV	.52	.63	.84
BP systolic	Rest	.80	.82	.78
	HV	.76	.71	.84
BP diastolic	Rest	.62	.88	.86
	HV	.76	.82	.84
Respiratory rate	Rest	.64	.82	.90
	HV	.38	.72	.85
Respiratory minute volume	Rest	.60	.80	.87
	HV	.62	.79	.88
O_2 uptake	Rest	.67	.80	.62
	HV	.70	.89	.88
Respiratory quotient	Rest	.60	.80	.69
	HV	.43	.67	.47

DISCUSSION

Whenever a test is repeated, adaptation takes place. The magnitude of this effect depends on the physiological parameter used and to a lesser extent on the nature of the stress test employed. In the cardiovascular system, heart rate and systolic blood pressure are especially affected by adaptation; in the respiratory system, the same is true for O_2 equivalent. In contrast with the results based on rest and strain scores, change scores (differences) show only modest adaptation effects. The magnitude of these depends largely on the type of stress test used. In order to avoid adaptation effects, repeated testing is therefore strongly recommended.

The magnitude of the stability coefficients depends on the parameter studied. This may be due to differences in tonic fluctuations among physiological parameters; that is, the non-systematic biological and situational variance exhibited by different measures (Myrtek, 1984). However, in contrast to the pattern reported in the literature, the stability coefficients of change scores were sometimes larger than the corresponding rest or strain scores in these studies. This implies that individual reactivity as represented by change scores may be a relatively stable characteristic. Comparison of different change scores (D, AHA and ALS) revealed no substantial differences in stability coefficients. The choice of an appropriate change score should therefore be based on other criteria.

Before applying a correction method that discards the simple difference between strain and rest scores, the investigator should consider whether a correction for change scores is truly worthwhile. The disadvantage of relating data points to regression lines (ALS) or axes (AHA) is that change scores become dependent on the subject sample from which the axes are calculated. Individual change scores are no longer comparable with change scores derived from different samples. Furthermore, the proper application of the ALS or AHA techniques requires that initial and final values be significantly correlated, that the regression of final values on initial values be linear, and that the variance along the regression line be evenly distributed (ie, homoscedastic). For these reasons, the use of single arithmetic differences is recommended.

The clinical diagnostic methods used at present assume that inferences about habitual responses may be made based on the results of a single examination or test. Our results for correlations of single measurements with habitual scores show that measures from a second test session are better predictors of the habitual values than are measures from the first session. Further examinations only slightly enhance the prediction of habitual response patterns.

In summary, we would recommend a second examination if this is possible. The results of a retest not only take into account adaptation effects but may also prove to be beneficial diagnostically.

REFERENCES

Amthauer R (1955) Intelligenz-Struktur-Test. Hogrefe, Göttingen

Lacey JI, Lacey BC (1962) The law of initial value in the longitudinal study of autonomic constitution: reproducibility of autonomic responses and response patterns over a four year interval. Ann NY Acad Sci 98: 1257-1290

Myrtek M, Foerster F, Wittman, W (1977) Das Ausgangswertproblem. Theoretische Überlegungen und empirische Untersuchungen. Zeitschrift für experimentelle und angewandte Psychologie 24: 463-491

Fahrenberg J, Walschburger P, Foerster F, Myrtek M, Müller, W (1983) An evaluation of trait, state, and reaction aspects of activation processes. Psychophysiology 20: 188-195

Myrtek M (1984) Constitutional Psychophysiology: Research in Review. Academic Press, New York

CHAPTER 9

The Quantification of Cardiovascular Reactivity in Longitudinal Studies

Jochen Fahrenberg, Hermann-Josef Schneider, Fritz Foerster, Michael Myrtek and Wolfgang Müller

Forschungsgruppe Psychophysiologie, Universität Freiburg, Belfortstrasse 20, D-7800 FREIBURG i.Br., Federal Republic of Germany

SUMMARY

Three methodological issues concerning the quantification of cardiovascular reactivity are presented in this paper: (1) the evaluation of change and range scores; (2) the stability of, and (3) the consistency and generalisability of selected cardiovascular measures. Conclusions and recommendations are based primarily on the findings of a recent multivariate laboratory-field study of 58 male students replicated at intervals of three weeks, three months and one year.

INTRODUCTION

Longitudinal studies concerned with cardiovascular risk and disease must acknowledge a number of basic questions and methodological issues that are critical to the assessment of cardiovascular reactivity. Table 1 outlines several principles underlying the selection and standardisation of parameters relevant to the assessment of reactivity. The present contribution is concerned with three methodological issues important in quantification of cardiovascular reactivity: (1) the evaluation of change scores and range scores; (2) the stability of, and (3) the consistency and generalisability of selected cardiovascular measures.

The findings are based primarily on a recent multivariate laboratory-field study conducted with 58 male students, which was replicated at intervals of three weeks, three months and one year. The present study is part of a larger research programme designed to evaluate individual differences in psychophysiological activation processes (Fahrenberg et al 1983; Fahrenberg and Foerster 1982; Foerster et al 1983; Myrtek 1984).

Figure 1 briefly illustrates the basic design used in the experiment and provides the context of the methodological discussion. The subjects were 58 male students of physical education. The study comprised psychological and psychophysiological assessments under various conditions: (1) at home (sleep recording of ECG); (2) in the Polygraph Laboratory while performing several tasks like Mental Arithmetic (MA), Reaction Time Test (RT), Free Speech (FS) and Cold Pressor (CP), interspersed with conditions of Rest (R1, R2), Relaxation (RE1, RE2) and Anticipation (AMA, AFS, ACP); (3) exercise tests employing a bicycle ergometer, which increased stepwise from 100 watt to maximal exertion (Emax); (4) monitoring under various pre-start conditions, including Relaxation Indoor (REI), Relaxation in the

Table 1. Suggested principles underlying the selection and standardisation of parameters suitable for the assessment of cardiovascular reactivity in longitudinal studies.

FUNDAMENTAL REQUIREMENTS	APPROPRIATE FORMS OF DATA ANALYSIS FOR THE ASSESSMENT OF THE CORRESPONDING REQUIREMENTS
1 Validity Relevance of measure for cardiovascular functions: discriminative efficiency between rest and challenge conditions known to provoke cardiovascular reactions; discriminative efficiency between subjects; discriminative efficiency between specific pathological conditions	Two-factorial ANOVA (subjects or groups of patients and conditions) with significant main effects for both subjects or groups of patients and conditions: multi-variate t-tests; discriminant analysis
2 Reliability Relatively small error variance	Error variance of ANOVA design; low frequency of missing data; relatively high short-term (eg, following 30 min interval) stability
3 Adequate scaling of observed reactivity Measurement of change, considering differences in initial value (baseline) as well as dynamic properties of cardiovascular regulation (synergisms)	Change scores (DIFF, ALS, AHA, Lykken); parameters from dynamic function tests (eg, half-time recovery, dampening, accuracy of regulation)
4 Stability (Reproducibility) Relative stability of rank order of individual differences between occasions (temporal stability across replications of the 'same' condition of measurement)	Correlation (R-technique across subjects), eg after 4 weeks, 12 months; 3-factorial ANOVA (subjects, conditions, replications), component attributable to trend in time series
5 Consistency Relative consistency of individual differences between fairly similar situations (trans-situational stability across various conditions of measurement, ie, tasks, challenges, stimulation)	Correlation (R-technique and intra-individual P-technique within subjects); partition of covariance with pooled estimates for particular aspects of covariation, eg, conditions (T-technique), profiles of variables (O-technique)

cont/

Table 1 (cont)

6 <u>Generalisability</u> Validity and reliability across populations (students, patients); validity and reliability across laboratory or clinic settings and field settings (daily life, at work)	ANOVA designs; correlation; coefficient of generalisability
FURTHER BASIC ASPECTS	
1 <u>Patterning</u> Contribution of a particular parameter to the recognition of cardiovascular patterns (synergisms, stimulus specific response patterns, dissociations of synergistic action);	Correlation; factor loading, pattern indices
recognition of individual specific response (idiosyncratic) patterns of possible aetiological importance (locus minoris resistentiae sive majoris irritationis/vulnerability)	Specificity analysis: components of ANOVA, MANOVA; concordance; O-Q-correlation; Mahalanobis-distances
2 <u>Aggregation</u> Contribution to more global assessment of cardiovascular conditions operationally defined by composite indices, eg, outflow of sympathetic nervous system, ergotropic phase	Correlation; item analysis; factor analysis; multiple regression designs
3 <u>Correlation</u> Functional relationships between various categories of physiological-neuroendocrinological variables; psychophysiological relationships between cardiovascular measures and psychological (subjectively reported, behavioural, psychosocial) variables - in order to represent a complementary pattern of human adaptation or maladaptation based on cardiovascular psychophysiology	Correlation (bivariate, multiple, canonical); pattern analysis; path analysis and LISREL
4 <u>Cost benefit</u> Evaluation and acceptability (eg inflicted discomfort, compliance) and utility (eg increment of validity per unit time and money; non-redundancy as compared to other methods)	Methodology developed in evaluation research

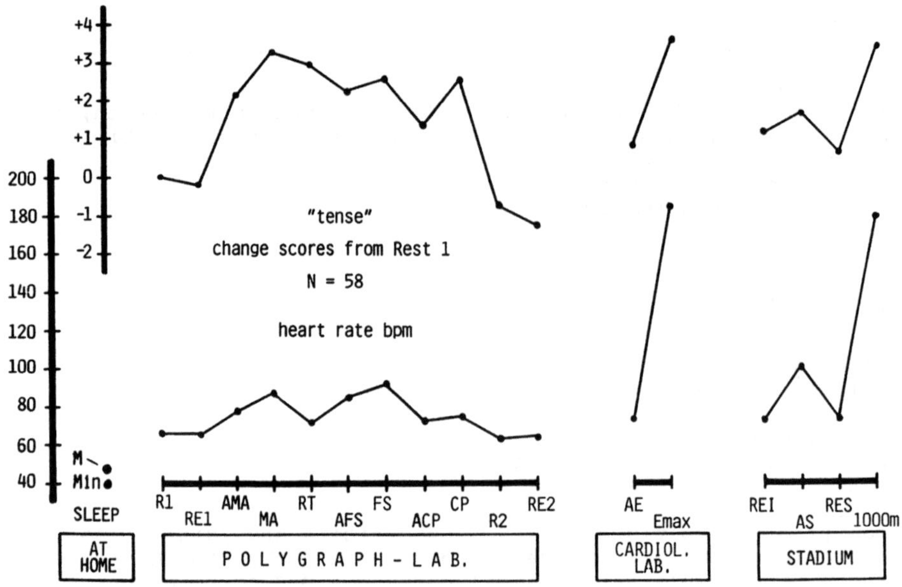

Fig 1. Time course of self-reported tenseness (change scores) and heart rate (bpm) across various experimental phases from sleep to 1,000m run. See text for abbreviations.

Stadium (RES), Approaching the Start (AS) and a 1,000 metre run (1000m). Figure 1 depicts the time course of self-reported activation and heart rate across these experimental phases from sleep to 1000m run. In this study 40 psychological and physiological parameters were employed to characterise the activation process.

The entire assessment was repeated after three weeks and again after three months. Recordings in the Polygraph Laboratory and sleep recording were also repeated with N=42 subjects after a one-year interval. A laboratory-field comparison was conducted to test whether or not individual differences in intensity and pattern of activation exhibited in certain field situations outside the laboratory can reliably be predicted from standardised assessment procedures in the laboratory. A number of psychophysiological hypotheses have been formulated and tested in this multivariate study (an extended report on our general research strategy and methods may be found together with the experimental results in Fahrenberg et al, 1984 and Foerster, 1984).

In our research, instead of using the problematic term 'stress', we prefer the traditional and more neutral term 'reactivity' and that of multivariate 'activation processes'. The study of activation processes, which are higher nervous system functions of varying intensity and synergistic patterning, requires a complementary psychological and physiological approach. Conceptually, a distinction can be made between psychophysiological activation as a tonic <u>state</u>, activation as an event-related <u>reaction</u>, and activation as a reactivity disposition or <u>trait</u>. Reactivity as a trait is thought to be a relatively stable and consistent disposition to exhibit higher degrees of activation state and/or reaction across specified stimulus conditions, with a response pattern more or less unique to each subject.

There are many possible theoretical orientations and research issues in cardiovascular psychophysiology which concern the momentary phasic reactions that fluctuate according to varying conditions of the organism (see Chapter 13). But with respect to predictive and longitudinal studies on risk and development of cardiovascular disease, the dispositional aspect describing the individual's habitual or constitutional reactivity or hyper-reactivity is the fundamental aspect of interest. To characterise individual differences in reactivity on cardiovascular measures a sufficient degree of stability is required. We believe stability coefficients of at least 0.70 (based on a three-week interval) are mandatory. Measures of individual differences that are hardly reproducible at retest under equivalent conditions of measurement are not suitable for this kind of research. To predict individual differences on a criterion measure from a different variable and/or another measurement condition, a correlation coefficient of at least 0.50 (that is 25% common variance) should exist between predictor and criterion.

It follows that the methodology used to study human adaptation would profit from the theory of assessment and prediction of individual differences developed and refined in differential psychology. Psychologists have always taken more interest in normal individual differences than have physiologists. Therefore, appropriate theoretical and statistical concepts are already available for use in differential cardiovascular psychophysiology.

CHANGE SCORES AND THE OPERATIONAL DEFINITION OF REACTIVITY

The development of an operational definition of reactivity requires the use of change scores. Myrtek (Chapter 8) has pointed out that in our laboratory three types of change scores, simple arithmetic differences (DIFF), autonomic lability scores (ALS), and distances from principal axis (AHA) have been used in several studies. The present experiment, extending from sleep recording to vita maxima during bicycle ergometer exercise as well as during the 1000m run, provides sufficient data to evaluate another type of change score, the so-called range-corrected change score proposed by Lykken (1966, 1972). In evaluating LYKKEN scores one should take into account several criteria of validity and reliability. Conclusions may differ depending on whether constitutional reactivity or momentary, phasic activation processes are considered. For our purpose, the relative stability of the change scores observed as compared to that of absolute raw scores (strain scores) and of the change scores DIFF, ALS and AHA is the primary interest.

A detailed analysis of heart rate and respiration rate reactions to various stimulus conditions of this experiment has been presented in a recent paper (Fahrenberg et al 1985). Results indicate that LYKKEN scores are generally less stable than raw measures. This is probably due to the elimination of individual differences in constitution from this transformation. But LYKKEN scores, for most cases, are also less stable than the simple arithmetic differences. Additional analyses indicated the following:

(1) DIFF 1 and LYKKEN 1 based on maximum range (over all conditions) correlated well with each other, as do DIFF 2 and LYKKEN 2 based on the intra-laboratory range.
(2) LYKKEN 1 scores are not adequately represented by LYKKEN 2 change scores.
(3) LYKKEN scores do not lead to a higher number of significant correlations among physiological parameters, or between physiological and psychological activation variables, than do other change scores.

The evaluation of various change score transformations with respect to adequate scaling of heart rate suggests the following conclusions:

(1) DIFF scores are sufficiently stable, concrete, computable for single subjects, and are generally adequate for use as change scores.
(2) ALS scores are less stable, theoretically derived, and are not computable for single subjects but require statistical analysis of a subject sample. However, they are useful in certain respects, eg, conservative tests of hypothetical specificities in patterning (see Fahrenberg et al 1983; Foerster et al 1983).
(3) AHA scores, similar to ALS, can be regarded as statistical abstractions, but are useful in analyses with respect to the problem of initial values when a separation of the spurious 'a (a-b)' effect is necessary (see Chapter 8).

Our study concerned heart and respiration rate, and it remains to be seen whether these findings generalise to the phasic activation of other physiological systems such as electrodermal activity. However, when habitual reactivity is assessed, raw measures of heart rate and DIFF change scores are recommended. Thus we would like to endorse the conclusions put forth by Myrtek in his independent study.

A quotation from Cronbach et al (1972) may serve to elucidate our point of view. Cronbach, who is a well-known proponent of psychometric theory, criticises DIFF change scores as follows: 'Although the unsuitability of such scores has long been discussed, they are still employed, even by some otherwise sophisticated investigators' (p334). Over the years we have tried to evaluate empirically more complex change scores, such as the ALS, AHA and LYKKEN scores. With increasing experience of such biometric analyses it appears to us that it might be more sophisticated to return to simple DIFF change scores. We recommend the use of raw measures (rest and strain scores) together with DIFF change scores to define an individual's reactivity.

Other types of change scores may be valuable for particular issues, for example, the investigation of the problems of initial value and regression to the mean, or may be essential for conservative statistical testing or response specificities. These are special applications that demand tailored data transformations.

RANGE MEASURES

The definition of an adequate baseline and the introduction of an initial rest period with reduced activation has always troubled psychophysiologists. In Freiburg, we have compared the initial and the final rest periods recorded in the Psychophysiology Laboratory with respect to various activation parameters. We found that the initial rest period is to be preferred, provided that a prior adaptation period is included. This usually consists of about 15 minutes for technical preparation followed by five minutes of baseline recordings.

However, our desire to obtain basal measures of parameters such as heart rate has persisted. We therefore decided to undertake sleep recording at the subject's home and to compare these measures with our laboratory and field recordings. This sleep recording was repeated one year later. The stability coefficient was 0.75 for heart rate (mean of entire night) and 0.55 for heart rate minimum (N=42).

The availability of an unusually extended range of measures from an individual's minimal heart rate during sleep to maximum heart rate during exercise motivated us firstly to look into the trends and stability of such range measures, and secondly to compute correlations among such range measures and a selection of other dispositional characteristics. Table 2 indicates that the range of heart rate, respiration rate and blood pressure observed in the Psychophysiology Laboratory tends to decrease across replications. The stability of such range statistics, with the exception of heart rate, appears to be rather low. The heart rate maximum range (Range 1 from sleep to 1000m run) is positively correlated with physical fitness. Range 2, based on the Psychophysiology Laboratory is not, however, correlated with Range 1 or heart rate measures during tasks other than Mental Arithmetic. Other

Table 2. Range measures of heart rate, respiration rate and blood pressure: means and coefficients of stability. Range 1 from minimum during sleep to maximum of 1000m run, blood pressure measured immediately afterwards. Range 2 from initial Rest phase to observed maximum within Psychophysiology Laboratory.

	Means				Stability coefficients 3wks	3mths	1year
	T1 N=58	T2 N=58	T3 N=58	T4 N=42	T 1/2 N=58	T 1/3 N=58	T 1/4 N=42
Heart rate							
Range 1	137.3	-	-	-	-	-	-
Range 2	32.8	27.3	25.2	23.5	.65	.62	.48
Respiration rate							
Range 1	37.7	-	-	-	-	-	-
Range 2	10.2	10.4	10.5	9.4	.36	.28	.32
Systolic blood pressure							
Range 2	45.9	43.9	41.2	32.5	.23	.26	.14
Diastolic blood pressure							
Range 2	31.6	32.8	32.6	27.6	.20	.29	.29

variables such as Type A behaviour, somatic complaints and adrenaline excretion do not provide much additional insight as to validity of heart rate range measures (see Fahrenberg et al 1984).

In most laboratory studies, extended range statistics are seldom available. The results obtained from this study do not, however, encourage such laborious examinations in conventional psychophysiological research. On the other hand, in some areas (eg, hypertension research), conventional assessments should be accompanied by additional information about basal levels (relaxation, sleep) and individual range to maximal exertion (see Chapter 10). A thorough account of 'reactivity' should take account of these biometric aspects, as well as attempt to define the relevant situational variables.

TRENDS AND STABILITY

Dispositional characteristics of cardiovascular regulation and risk should be evaluated with reference to normative data obtained from ample samples of subjects and patients. In addition, an adequate selection of variables is needed, together with different stimulus conditions and replications after short- and long-term intervals. In addition to the statistical issues discussed above, trends and stability coefficients have been computed for the variables employed in our laboratory-field study. A selection of these results for cardiovascular and respiratory measures is summarised in Table 3.

Table 3. Trends and coefficients of stability for selected variables recorded during various situations at intervals of three weeks (1/2), three months (1/3) and one year (1/4). REST initial resting condition, MA mental arithmetic, CP cold pressor, E100 and Emax bicycle ergometer 100 watt and maximum exercise, REL relaxation pre-start, 1000m run. Trends are estimated from a two-factorial ANOVA (subjects, occasions) in percent of total variance; coefficients of stability are given for raw measures as well as change scores (DIFF). Since the ergometer test and 1000m run were not repeated after a one year interval, these coefficients are not available.

		TRENDS		STABILITY COEFFICIENTS					
		ANOVA %		raw scores			change scores (DIFF)		
Variable	Situation	1/3	1/4	1/2	1/3	1/4	1/2	1/3	1/4
Heart rate	REST	1	1	64	34	47	-	-	-
	MA	7	4	76	52	57	63	53	43
	CP	3	2	75	67	64	73	59	57
	E100	0	-	76	70	-	64	57	-
	Emax	2	-	73	65	-	69	58	-
	REL	2	-	57	65	-	09	16	-
	1000m	0	-	58	32	-	32	24	-

cont/

Table 3 (cont)

Variable	Situation	TRENDS ANOVA % 3T	4T	STABILITY COEFFICIENTS raw scores 1/2	1/3	1/4	change scores (DIFF) 1/2	1/3	1/4
Cardiac output (impedance technique)	REST	0	1	18	22	07	-	-	-
	MA	0	2	32	21	10	45	27	05
	CP	0	0	13	17	04	04	23	-37
Systolic blood pressure	REST	3	2	28	26	18	-	-	-
	MA	0	2	16	08	28	28	-01	19
	CP	9	8	36	27	10	38	24	-03
	E100	3	-	45	51	-	24	18	-
	Emax	0	-	42	31	-	36	21	-
	REL	2	-	30	29	-	15	15	-
	1000m	0	-	56	52	-	38	38	-
Diastolic blood pressure	REST	1	4	32	31	34	-	-	-
	MA	0	0	24	42	35	26	38	42
	CP	7	2	20	41	27	19	19	03
	E100	8	-	38	39	-	02	19	-
	Emax	3	-	44	26	-	27	17	-
	REL	11	-	26	40	-	41	-15	-
	1000m	10	-	44	37	-	39	31	-
Pulse wave velocity	REST	1	0	63	56	58	-	-	-
	MA	7	4	69	66	37	24	35	03
	CP	5	2	61	66	56	43	23	19
Respiration rate	REST	3	2	66	55	65	-	-	-
	MA	1	1	64	58	51	60	47	34
	CP	1	1	83	63	69	52	36	45
	E100	0	-	74	67	-	53	52	-
	Emax	4	-	73	64	-	69	68	-
	REL	1	-	72	79	-	01	-05	-
	1000m	1	-	66	63	-	62	64	-
Respiratory minute volume	REST	0	-	31	41	-	-	-	-
	E100	0	-	34	22	-	46	35	-
	Emax	1	-	74	63	-	73	66	-
Oxygen uptake	REST	6	-	12	31	-	-	-	-
	E100	10	-	10	00	-	07	00	-
	Emax	5	-	69	61	-	72	67	-
Respiratory quotient	REST	11	-	-08	03	-	-	-	-
	E100	16	-	10	22	-	07	45	-
	Emax	15	-	-09	26	-	16	20	-

The general trend component seems to make up only a relatively small proportion of total variance. The coefficients of stability after an interval of three weeks are fairly satisfactory for heart rate, pulse wave velocity and respiration rate. For the other variables, a further decrease of stability coefficients is evident over a period of three months to one year. For many variables, change scores (DIFF) are unstable, and with the exception of heart rate are insufficient for reliable estimation of habitual individual differences. These general conclusions are of course influenced by the particular situation of measurement (high coefficients for example for maximum exercise), possible problems of measurement, and the degree of instrumental reliability. The latter is remarkably high for most of these variables, as shown by short-term stability coefficients (Fahrenberg 1983). The reported stability coefficients for retests of three weeks and three months intervals indicate a considerable amount of fluctuation in activity, thus limiting empirical prediction of individual differences across various situations.

CONSISTENCY AND GENERALISABILITY

The consistency and generalisability of heart rate and other activation parameters were studied by correlating data across various laboratory and field situations. Additionally, a generalisability study (G-study) as described by Cronbach et al (1972) was performed.

The correlation coefficients in Table 4 indicate the amount of cross-situational consistency in heart rate, respiration rate, systolic and diastolic blood pressure for a number of experimental situations. The data are abstracted from the initial examination of the study outlined earlier. For heart rate the coefficients obtained are modest to low. However, substantial correlation coefficients of r=0.50 or more are evident between the various Rest and Relaxation conditions; within the various laboratory conditions such as Mental Arithmetic, Anticipation of Free Speech, and the Cold Pressor; and within the field pre-start situations.

Heart rate measured during the 1000m run could not be reliably predicted from data recorded during ergometer exercise. Consistency between laboratory and field data, generally, is less strong than for data recorded within laboratory situations. The blood pressure and respiration rate data exhibit the same pattern, but the coefficients are generally even lower than for heart rate.

The generalisability study was performed for psychological and physiological parameters based on raw (strain) scores as well as DIFF change scores (see Fahrenberg et al 1984). Table 5 depicts a selection of results. Psychological data (changes in mood ratings) and physiological data (strain scores) were recorded in ten experimental conditions: Mental Arithmetic, Reaction Time Test, Anticipation of Free Speech, Cold Pressor Test, Anticipation of Ergometry, Ergometry maximal exercise, Relaxation indoor, Approaching Stadium, Relaxation Stadium, and 1000m run.

The estimated coefficients of generalisability obviously depend on the relationship between variance components, the proportion of variance due to Subjects compared with variance due to Conditions being of primary importance. This relationship can be manipulated by selecting or eliminating conditions which contribute to the

Table 4. <u>Trans-situational consistency</u>. Coefficients of correlation between various laboratory and field situations in 58 male sport students. Above the diagonal, systolic and diastolic blood pressure (SBP, DBP in mmHg) are shown. Below the diagonal, heart rate (HR in bpm) and respiration rate (RR in cycles per minute) are detailed.

		1	2	3	4	5	6	7	8	9	10	11	12	13
1	Sleep	HR	SBP	-	-	-	-	-	-	-	-	-	-	-
	minimum	RR	DBP	-	-	-	-	-	-	-	-	-	-	-
2	Sleep	95		-	-	-	-	-	-	-	-	-	-	-
	mean	-		-	-	-	-	-	-	-	-	-	-	-
3	Rest 1	29	34		69	66	66	73	34	19	34	33	25	26
	(Lab)	-	29		48	63	56	45	09	11	31	38	24	19
4	Mental	15	22	55		73	76	81	25	-01	30	29	26	07
	arithmetic	-	24	30		59	66	64	01	17	25	24	27	12
5	Anticipation	10	19	63	70		82	74	24	01	10	22	24	04
	free speech	-	31	63	57		61	47	10	20	33	23	28	28
6	Cold pressor	28	31	61	57	50		80	25	10	09	17	13	10
	(1+2 min)	-	12	43	38	58		62	17	12	14	11	20	00
7	Relaxation 2	34	40	84	60	52	65		32	-01	33	39	36	13
	(Lab)	-	19	67	33	56	48		16	29	21	16	37	-01
8	Ergometer	40	45	42	56	33	57	49		45	20	33	29	16
	100 watt	-	-02	28	00	20	19	23		62	28	19	34	21
9	Ergometer	23	32	22	39	25	38	32	52		12	17	07	32
	maximum	-	01	08	17	09	13	04	51		22	22	37	32
10	Relaxation	53	56	50	44	34	53	54	46	43		63	57	37
	indoor	-	39	75	22	55	48	66	37	-01		56	43	37
11	Approaching	39	46	41	49	42	54	44	58	54	76		66	39
	stadium	-	29	34	13	42	25	17	09	15	27		46	24
12	Relaxation	52	56	54	51	37	52	55	48	40	93	77		26
	pre-start	-	38	76	20	60	47	54	21	-05	83	34		29
13	1000m run	31	39	04	04	-05	19	16	30	35	45	43	46	
		-	-01	00	-15	-05	-03	02	44	38	00	-17	-01	

Table 5. Generalisability study employing four psychological and four physiological activation parameters. Table 5a is based on ten laboratory and field conditions repeated on three occasions (trials) at intervals of three weeks and three months. Table 5b is based on eight laboratory and field situations excluding the conditions of maximal exertion, ie Ergometer and 1000m run. Components of variance for Subjects (S), Conditions (C), the Subjects by Conditions Interaction (SxC), Trials within Subjects (TwS) and Residual (R) are given in percent of total variance. Generalisability coefficients (estimated coefficients p^2) for various combinations of Conditions and Trials (1/1, 1/10, 10/1 and 5/5) are also shown.

(a) Ten conditions	10 Conditions, 3 Trials. Components of variance (%) S	C	SxC	TwS	R	Generalisability coefficient $E(p^2)$ 1C 1T	1C 10T	10C 1T	5C 5T
Tense	10	35	14	3	38	.10	.16	.45	.45
Active	15	27	17	5	36	.15	.24	.53	.57
Motivated	16	14	13	6	51	.16	.33	.53	.65
Irritated	20	13	18	3	46	.20	.36	.65	.70
Heart rate	3	92	2	2	1	.03	.03	.22	.14
Respiration rate	4	80	9	1	6	.04	.04	.27	.18
Systolic blood pressure	10	46	10	6	28	.10	.15	.41	.43
Diastolic blood pressure	15	27	12	8	38	.15	.26	.49	.58

(b) Eight conditions	8 Conditions, 3 Trials. Components of variance (%) S	C	SxC	TwS	R	Generalisability coefficient $E(p^2)$ 1C 1T	1C 10T	10C 1T	5C 5T
Tense	11	28	15	4	42	.11	.19	.46	.49
Active	16	21	18	6	40	.16	.26	.52	.59
Motivated	15	13	12	8	52	.15	.32	.49	.63
Irritated	18	16	21	5	41	.18	.30	.59	.64
Heart rate	29	37	11	9	15	.29	.37	.66	.71
Respiration rate	26	25	19	3	28	.26	.35	.71	.71
Systolic blood pressure	17	31	6	15	30	.17	.28	.43	.58
Diastolic blood pressure	18	24	10	11	37	.18	.32	.51	.64

variance component 'Conditions'. Therefore, an additional G-study without the two conditions of maximal physical exertion was computed to demonstrate the increase in $E(p^2)$. These estimated generalisability coefficients are presented for several combinations of one or many conditions and/or trials (1C/1T, 1C/10T, 10C/1T, 5C/5T).

After excluding the two aforementioned conditions, generalisability coefficients (Ep^2) for psychological and physiological parameters are roughly in the same range, and are highest for heart rate and respiration rate. But as soon as conditions of intense strain are included in the analysis, the value of $E(p^2)$ for physiological parameters (especially for heart rate and respiration rate) sharply decrease, while the coefficients for psychological variables remain fairly stable.

Generalisability studies, as described by Cronbach et al (1972), may be helpful in planning experimental designs, in order to achieve maximal information with limited resources of time and money by employing either more experimental conditions or more trials. In evaluating these results, it should be noted that the analyses were performed on data from a particular, although relatively broad set of situations in a rather homogeneous sample of male students. Results may differ for more heterogeneous samples. Unfortunately, no reports of this type are available in the literature, so comparison is ruled out. The present study contributes to a more sophisticated approach to the evaluation of temporal and cross-situational consistency of activation parameters. As such, it addresses a fundamental methodological issue in selecting cardiovascular measures suited to the evaluation of individual differences in reactivity in longitudinal studies of risk and development of disease.

CONCLUSIONS

The following five points serve to summarise the present findings:

(1) Quantification of cardiovascular reactivity in longitudinal and risk studies of human adaptation will profit from the methodology developed in differential psychology, which includes the theoretical and statistical concepts of assessment and prediction.

(2) In addition to the validity of a given cardiovascular parameter and the actual reliability of its measurement, moderate to high stability of an individual's response characteristics is essential for the assessment of differences in dispositional reactivity.

(3) There have been several types of change score proposed for the scaling of heart rate and other physiological functions. Based on a relatively broad evaluation of such change scores, we would like to recommend simple arithmetic differences between rest and strain scores, together with strain scores, to assess reactivity. The use of other types of change scores should be restricted to special issues.

(4) The stability of individual differences in heart rate, and its consistency and generalisability across various situations, is relatively high as compared with other parameters under study. However, even with heart rate measures, consistency across a one-year interval and from laboratory to field settings is hardly sufficiently reliable to predict individual differences in reactivity.

(5) In differential psychophysiology, such vague trait concepts as 'psychophysiological reactivity' or

'cardiovascular reactivity' have not been sufficiently substantiated by empirical finding (see Fahrenberg 1983). Such global concepts should therefore be replaced by a more precise terminology with respect to distinct cardiovascular parameters. This should in turn indicate a relatively high degree of temporal and trans-situational consistency.

REFERENCES

Cronbach LJ, Gleser GC, Nanda H, Rajaratnam N (1972) The Dependability of Behavioral Measurements: theory of generalizability for scores and profiles. Wiley, New York

Fahrenberg J (1983) Psychophysiologische Methodik. In: Groffmann KJ, Michel L (eds) Enzyklopädie der Psychologie. Psychologische Diagnostik. Band 4 Verhaltensdiagnostik. Hogrefe, Göttingen, p 1-192

Fahrenberg J, Walschburger P, Foerster F, Myrtek M, Müller W (1983) An evaluation of trait, state and reaction aspects of activation processes. Psychophysiology 20: 188-195

Fahrenberg J, Foerster F, Schneider HJ, Müller W, Myrtek M (1984) Aktivierungsforschung im Labor-Feld-Vergleich. Minerva, München

Fahrenberg J, Foerster F, Schneider HJ, Müller W, Myrtek M (1985) Adequate scaling of heart rate reactions. In: Orlebeke JF, Mulder G, van Doornen LJP (eds) Psychophysiology of Cardiovascular Control. Plenum Press, New York

Fahrenberg J, Foerster F (1982) Covariation and consistency of activation parameters. Biol Psychol 15: 151-169

Foerster F (1984) Computer-Unterprogramme zur Biosignal-Analyse. Springer, Heidelberg

Foerster F, Schneider HJ, Walschburger P (1983) Psychophysiologische Reaktionsmuster. Minerva, München

Lykken DT (1972) Range correction applied to heart rate and to GSR data. Psychophysiology 9: 373-379

Lykken DT, Rose R, Luther B, Maley M (1966) Correcting psychophysiological measures for individual differences in range. Psychol Bull 66: 481-484

Myrtek M (1984) Constitutional Psychophysiology. Academic Press, New York

CHAPTER 10

The Measurement of Cardiac Reactivity in Clinical Populations

A Perski, K Orth-Gomér, A Hamsten, M Roll, U de Faire, T Theorell and S Zetterquist

National Institute for Psychosocial Factors and Health, Karolinska Institutet, Box 60210, S-104 01 STOCKHOLM, Sweden
and
Departments of Internal Medicine and Clinical Physiology, Danderyd Hospital, Karolinska Institutet

SUMMARY

Unique methodological issues arise in the study of cardiac reactivity as a predictor of morbidity and mortality in clinical populations. One such issue concerns the considerable clinical and physiological difference between patients that has to be taken into account when evaluating cardiac reactivity. Various range correction techniques are described, including a technique which expresses the cardiac reaction as a percentage of the peak level obtained during maximal effort. Furthermore, new parameters of cardiac reactivity are proposed. These include measurements of recovery time after challenge, inability to relax, frequency and type of ventricular arrhythmias as well as ST-segment changes.

CARDIAC REACTIVITY IN CORONARY PATIENTS

The issue of cardiac reactivity has been studied intensively by psychophysiologists over the last three decades (Krantz and Manuck 1984). The development of such concepts as individual response and stimulus response specificity gradually modified the traditional notion that cardiac reactivity is an expression of general arousal. Instead, the specificity concept implied a new type of relationship between cardiac reactivity, symptoms and disease development, wherein a correlation is to be found between the physiological parameter in which the patient shows his or her maximal reactivity, and the symptom which characterises the medical problem. Thus, the aetiology of essential hypertension, myocardial infarction, stroke and sudden cardiac death could possibly involve individual response specificity in the cardiovascular system. Unfortunately, despite a large body of research on cardiac reactivity (Fahrenberg et al 1983), only a few studies have actually been conducted to test this notion. One study attempted to link individual response specificity to the development of essential hypertension (Engel and Bickford 1961), and another to development of coronary heart disease (Keys et al 1971). More recently, studies have been conducted on the role of the Type A behaviour pattern and its cardiovascular reactivity correlates in the aetiology of myocardial infarction (Glass et al 1980; Krantz and Durel 1983). The results of these studies are encouraging. However, we are convinced that the only adequate way of resolving the methodological issues which arise from the evaluation of reactivity as a predictor of disease development is to conduct a series of prospective studies. Only then can the problems of what kind of physiological parameters and provocations to use, and how to analyse the data, be resolved.

Equally important as the relationship between reactivity and disease development is the role of cardiovascular reactivity in prognosis after myocardial infarction. Excessive reactivity coupled with a damaged myocardium may constitute a major risk factor in the development of new infarctions or sudden cardiac death. Cardiovascular reactivity could increase the risk of fatal episodes by imposing sudden excessive oxygen demand on the damaged myocardium by means of increases in heart rate, systolic blood pressure and myocardial contractility. The mortality risk may also increase through spasm in the coronary blood vessels. Finally, heightened cardiac reactivity may decrease the threshold for potentially fatal ventricular arrhythmias. As far as we are aware, no prospective study has evaluated cardiac reactivity in such a context. It is therefore proposed to discuss some methodological issues which may be important in such studies.

RESPONSE MAGNITUDE AND RANGE CORRECTION

The first concern is with the evaluation of cardiovascular response magnitude. Studies on clinical populations differ considerably from those conducted on healthy subjects (see Chapter 9). Different degrees of cardiac damage set limits on the ability of patients to respond to provocations. The frequent use of pharmacological treatment in coronary populations also poses constraints on response capability. Consequently, evoked cardiovascular responses have a different meaning that depends on the range of possible responses. A small heart rate response may vary in significance according to whether the heart rate level that produces myocardial ischaemia is low or high. In order to deal with this issue, it is necessary to introduce range correction techniques that take into account the individual physiological framework of the patient.

Lykken (1972) introduced the concept of range correction, which involved a division of the response (the difference score between rest and level during provocation) by the individual's total range. In order to illustrate this kind of range correction technique and to introduce additional methods, we will use the results from a study recently conducted in our laboratory on 20 young post-myocardial infarction patients (age below 45 years). Of these 20 patients, 12 were treated with a β_1-selective blocking agent and eight were without ß-blockade. These patients were subjected to an extensive cardiac reactivity testing procedure. They were given a series of psychomotor tests (reaction time task, computer squash game, mental arithmetic test, Stroop colour-word test), emotional tests (questionnaires, stress interview), cold pressor test and a maximal work stress test on a bicycle ergometer. Long periods of progressive relaxation were interposed between tasks. Figure 1 illustrates the procedure, together with the heart rate and systolic and diastolic blood pressure responses of one untreated patient. After laboratory testing, a 24-hour Holter recording of ECG was obtained. In addition to the post-myocardial infarction patient group, a group of 20 young cardiac neurotic patients and a group of 19 healthy subjects matched to the post-myocardial infarction patients by age, occupation and geographic area, were subjected to the same procedure.

Table 1 illustrates the range correction techniques which have been used in the analysis of cardiovascular reactivity. The traditional method of analysis involves the computation of change scores from

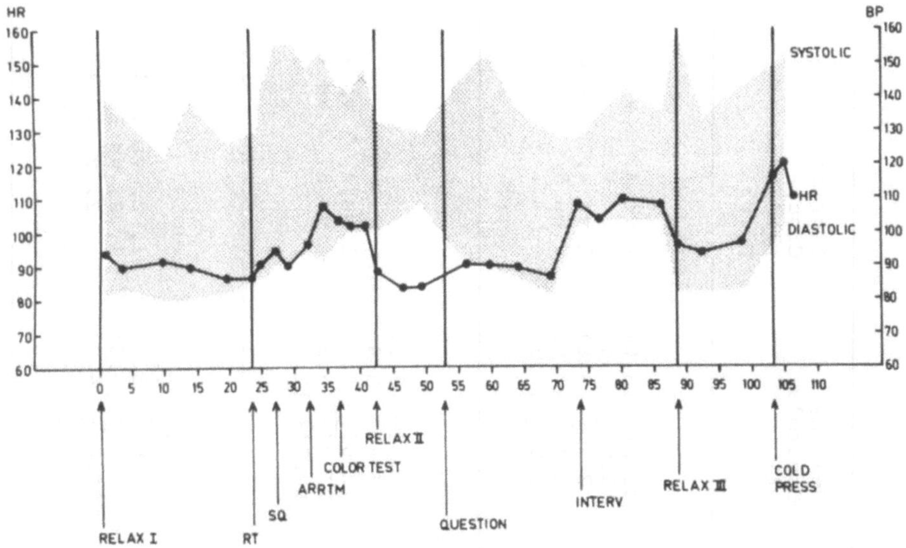

Fig 1. Cardiovascular reactions of a young post-myocardial infarction patient to laboratory challenges. RELAX = progressive relaxation, RT = reaction time test, SQ = squash computer game, ARRTM = mental arithmetic test, QUESTION = mood and exhaustion questionnaires, INTERV = stress interview, COLD PRESS = cold pressor test, COLOR TEST = Stroop colour test.

the baseline immmediately preceding task to a level obtained during task. We obtained change scores for heart rate responses which are presented in Table 1a. This analysis did not show any differences in heart rate responses between untreated post-myocardial infarction patients and cardiac neurotic or healthy subjects, although a difference occurred between patients treated with β_1-blockade and control groups. A range correction technique similar to that proposed by Lykken was applied to these data. Change scores were divided by the range of heart rate obtained on subtracting the lowest heart rate during sleep from the heart rate during maximal work load. The results are summarised in Table 1b, and again few differences in heart rate response are reliable.

We suggest that calculations which involve 'baseline' values always present a problem. In the case of traditional change scores, the immediate baseline level preceding tasks is itself influenced by the ability of the subject to relax in the often threatening laboratory conditions of the hospital. This is especially apparent when a series of tasks are administered. Consequently, reactive subject will not only show large reactions but their baseline levels may also be high, so change scores will under-estimate the true response. As the range correction technique proposed by Lykken also involves baseline levels, it shares this problem. A range correction technique was therefore developed that was free from the baseline problems, and put the reactions of patients into a meaningful clinical and physiological context. Heart rate levels obtained during different provocation tests were expressed as a percentage of the maximal heart rate obtained during a work stress

Table 1. Heart rate reactivity to different tests in four groups of subjects. Peak responses in the tests as assessed by change scores (a), Lykken range correction scores (b) and percentage of maximal heart rate (c). Means and standard deviations.

Experimental Condition	Post-MI No ß-blockade (1)	Post-MI ß-blockade (2)	Cardiac Neurotics (3)	Healthy Controls (4)	Significant Group Differences p<.05
		(a) Change scores			
Psychomotor	19.0(± 7.7)	13.4(± 5.0)	23.7(± 10.3)	23.8(± 10.6)	2 vs 3, 2 vs 4
Emotion	10.7(± 9.7)	8.8(± 3.5)	10.8(± 5.8)	8.8(± 3.5)	
Cold pressor	10.1(± 8.3)	5.8(± 4.9)	11.0(± 7.1)	2.7(± 3.6)	1 vs 4, 2 vs 4
Work 40 W	28.4(± 9.9)	25.0(± 9.4)	35.1(± 8.3)	33.3)± 7.9)	2 vs 3
Max 24 hour	75.1(± 13.1)	77.1(± 31.1)	83.9(± 12.7)	75.0(± 14.3)	
		(b) Lykken range correction scores			
Psychomotor	20.5(± 9.7)	17.2(± 5.4)	18.3(± 7.8)	19.9(± 9.2)	
Emotion	12.1(± 12.8)	12.4(± 6.5)	8.5(± 5.1)	7.4(± 3.9)	
Cold pressor	10.8(± 7.8)	7.2(± 7.3)	8.6(± 5.7)	2.6(± 3.1)	1 vs 4, 3 vs 4
Work 40 W	29.7(± 10.1)	33.4(± 9.9)	26.9(± 4.8)	27.2(± 6.9)	2 vs 3, 2 vs 4
Max 24 hour	78.3(± 9.7)	92.5(± 17.2)	64.9(± 10.7)	62.9(± 12.4)	
		(c) Percentage of maximal heart rate			
Psychomotor	59.9(± 11.5)	57.5(± 12.9)	51.4(± 8.6)	52.1(± 7.9)	
Emotion	53.9(± 13.2)	53.5(± 13.5)	43.6(± 7.4)	43.8(± 5.8)	1, 2, vs 3, 4
Cold pressor	56.8(± 13.8)	56.3(± 12.9)	45.9(± 7.1)	42.9(± 5.6)	1, 2, vs 3, 4
Work 40 W	66.5(± 14.0)	65.8(± 12.4)	54.8(± 8.6)	54.8(± 4.9)	1, 2, vs 3, 4
Max 24 hour	86.0(± 7.1)	94.9(± 10.3)	74.3(± 8.2)	74.1(± 9.0)	1, 2, vs 3, 4

test. The results in Table 1c show a clear dissociation between the post-myocardial infarction patient group and the comparison groups. The method also diminishes the effect of pharmacological treatment, which otherwise affects both baseline and maximal heart rate levels to a considerable extent.

It has repeatedly been shown that there is a low correlation between responses to different provocation tests in the laboratory, and between laboratory measurements and 24-hour ambulatory recordings. Table 2a shows the correlations obtained in the present study, both for change scores and for reactions expressed as a percentage of maximal heart rate. Correlations between change scores in the laboratory varied between .30 to .41. When our range correction technique was applied, correlations of .85 to .95 were obtained. A dramatic rise in the correlation coefficients was also seen when laboratory and real-life heart rate data were compared (Table 2b). Additional analyses indicated that absolute heart rate values were better predictors of heart rate obtained in other situations than were change scores.

Table 2. Product-moment correlations between heart rate responses as expressed by change scores from baseline and percentages of maximal heart rate (in parenthesis), (a) correlations between different laboratory challenges, (b) correlations between laboratory and 24-hour real-life recordings

(a)	Psychomotor	Emotional	Cold pressor
Emotional	.40 (.92)		
Cold pressor	.41 (.85)	.35 (.86)	
Work 40 W	.38 (.95)	.37 (.94)	.54 (.92)

(b)	Psychomotor	Emotional	Cold pressor	Work 40 W
Maximal HR 24-hour period	.16 (.43)	-.08 (.59)	.04 (.60)	.33 (.55)
Maximal HR during activity, 24-hour period	.11 (.48)	.05 (.66)	.07 (.68)	.26 (.59)

It should be emphasised that the decision regarding choice of technique for data analysis can only be made adequately in prospective studies. Since such studies have not yet been conducted, the arguments in favour of this range correction technique should be considered as inferential rather than conclusive.

OTHER PARAMETERS OF CARDIAC REACTIVITY

The final methodological issue that deserves mention concerns the nature of the data collected. Prospective studies on clinical populations suggest that parameters other than heart rate and blood pressure are required in the analysis of cardiovascular reactivity. Table 3 summarises the parameters that might be valuable as additional indicators of autonomic nervous system influences on cardiovascular function. The basic rationale for proposing such parameters is the realisation that different patient groups have unique cardiovascular limitations. The demonstration of the effects of cardiac reactivity on survival has therefore to be evaluated in the appropriate physiological and clinical context. More attention needs to be paid to the use of stimuli or situations which are effective in the provocation of responses (see Chapters 1, 2 and 3). The development of more realistic and individually meaningful conditions will maximise the possibility of documenting the importance of cardiac reactivity.

Table 3. Parameters for inclusion in prospective studies of cardiac reactivity as a predictor of survival after myocardial infarction

1. Ability to relax during day or night.
2. Recovery time after physical or emotional challenge.
3. Reactions to individually meaningful stimuli or stimuli specifically important for the clinical population.
4. Parameters uniquely relevant to the clinical population:
   (a) Rate-pressure product rather than HR and blood pressure separately.
   (b) Frequency and type of ventricular arrhythmias.
   (c) ST-segment depressions in ECG.

REFERENCES

Engel BT, Bickford AF (1961) Response specificity. Arch Gen Psychiat 5: 478-489

Fahrenberg J, Walschburger P, Foerster F, Myrtek M, Müller W (1983) An evaluation of trait, state and reaction aspects of activation processes. Psychophysiology 20: 188-195

Glass DC, Krakoff LR, Contrada R (1980) Effects of harassment and competition upon cardiovascular plasma catecholamine responses in Type A and Type B individuals. Psychophysiology 17: 453-463

Keys A, Taylor HL, Blackburn H, Brozek J, Andersson JT, Somonson E (1971) Mortality and coronary heart disease among men studied for 23 years. Arch Int Med 128: 201-214

Krantz DW, Durel LA (1983) Psychobiological substrates of the Type A behavior pattern. Health Psychology 2: 393-411

Krantz DS, Manuck SB (1984) Acute psychophysiological reactivity and risk of cardiovascular disease: a review and methodologic critique. Psychol Bull 96: 435-464

Lykken DT (1972) Range correction applied to heart rate and to GSR data. Psychophysiology 9: 373-379

CHAPTER 11

The Effects of Antihypertensive Medication and its Withdrawal on Psychophysiological Examinations

Heinz Rüddel

Medizinische Universitätsklinik, Sigmund-Freud-Strasse 25, D-5300 BONN 1, Federal Republic of Germany

SUMMARY

The problems associated with stress testing of patients undergoing antihypertensive therapy are described. The withdrawal syndrome following abrupt discontinuation of therapy may include reflex blood pressure increases, adrenoreceptor changes, and other physiological phenomena, together with a variety of symptoms. These effects have been observed after withdrawal from ß-blockers, clonidine and other pharmacological agents. If stress testing has to involve chronically medicated patients, it is suggested that the studies are carried out either after drug withdrawal for four weeks, or while patients continue with their medications.

INTRODUCTION

No studies have been published that have specifically examined the impact of discontinuing effective antihypertensive therapy on psychophysiological stress testing. It is difficult therefore to give exact recommendations on when to test hypertensive patients; should they be examined while still medicated, a few days after withdrawal from therapy, or several weeks after cessation of therapy? This chapter summarises data concerning withdrawal phenomena, and the recommendations on time course of testing are based on the pathophysiological responses induced by drug withdrawal.

WITHDRAWAL FROM BETA-ADRENERGIC BLOCKADE

Symptoms such as palpitations, tremor and sweating have been reported after abrupt withdrawal from propranolol, and appear to be associated with rebound hypertension (Hjemdahl and Olsson 1982). However, the incidence of these effects seems to be low in hypertensives in comparison to patients with coronary heart disease (Shand 1975; Garbus et al 1979). The mechanisms responsible for rebound phenomena after withdrawal of long-term ß-adrenoceptor blockade have not been fully clarified (Gilfrich and Rahn 1981). It would appear that the up-regulation of ß-adrenoceptors during long-term treatment seems to be important in the acute withdrawal syndrome (Boudoulas et al 1977; Maling and Dollery 1979). Increased aggregability of platelets, changes in the oxygen dissociation curve (Krämer and Kübler 1984), and the rebound increase in renin (Garrett and Kaplan 1980) and tri-iodothyramine (Kristensen et al 1978) have also been implicated. Although rebound hypertension is only seen in some patients, it has not yet proved possible to identify these patients beforehand.

The propranolol withdrawal syndrome is probably not restricted to treatment with this agent. Although Bolli et al (1981) did not observe hypersensitivity of ß-adrenoceptors after abrupt withdrawal of long-term therapy with slow release oxprenolol, an increased ß-receptor sensitivity and the clinical symptoms of rebound were reported after withdrawal from metaprolol and other ß-blockers (Petersen et al 1979; Bengtsson et al 1981; Ragno and Langlois 1982). It is therefore likely that most, if not all ß-adrenoceptor blocking agents will cause rebound phenomena in some patients, when they have been administered at an adequate dose.

REBOUND EFFECTS AND OTHER CARDIOVASCULAR DRUGS

Many cardiovascular drugs lead to reflex and end-organ receptor readjustments without rebound phenomena when therapy is withdrawn abruptly. In order to produce a rebound, the duration of drug action must be shorter than the readjustment time for the homeostatic mechanisms activated during drug administration (Gerber and Nies 1979). This may explain why individual cases of rebound hypertension have been observed after abrupt cessation of virtually all antihypertensive drugs with short half-lives. The plasma elimination half-lives of the most widely used antihypertensive drugs are summarised in Table 1 (based on Avery 1980). The table presents a simplified picture, because some drugs have active

Table 1. Summary of half-lives of drug elimination from plasma and intervals between doses (in hours) for common antihypertensive medications

	Elimination half-life	Typical intervals between doses
<u>Beta-blockers</u>		
Acebutolol	8-13	12-24
Atenolol	6-9	24
Propranolol	2-6	6-12
<u>Diuretics</u>		
Hydrochlorothiazide	2-15	12-24
Calcium antagonists	2-14	4-24
Clonidine	13	8-12
Captopril	4	6-12
Prazosin	2-5	8-12
Hydrallazine	2-4	8-12
Alpha-methyl-dopa	8	8-12
Reserpine	46-168	24
Guanethidine	120-240	24

metabolites with quite different half-lives (eg, acebutolol), and the half-lives of many different drugs of the same category have been grouped together (eg, the calcium antagonists verapramil, nifedipine, nitrendiprine and diltiazem).

Withdrawal from centrally-acting antihypertensive agents has also been studied extensively. By the early 1960s, there were reports of blood pressure rebounds together with symptoms such as insomnia, agitation, headache and nausea. These appear during the first few days after abrupt withdrawal from therapy with alpha-methyl-dopa (Garbus et al 1979). Similar alterations were described after sudden clonidine withdrawal (Goldberg et al 1977; Geyskes et al 1979). Some researchers have suggested that an increase in plasma catecholamines and up-regulation of alpha$_1$ adrenoceptors mediate the withdrawal syndrome after clonidine therapy (Whitsett 1983). No reports of captopril withdrawal symptoms in hypertensives have yet been published. However, patients with heart failure show increases in plasma angiotensin II, cortisol, plasma noradrenaline and blood pressure without alterations of cardiac output or electrolyte haemostasis after cessation of treatment with captopril (Maslowski et al 1981).

These observations clearly indicate that withdrawal symptoms may occur after cessation of virtually every effective antihypertensive therapy, although the incidence is low. Rebound phenomena of sympathetic readjustment are common, may persist for two to three weeks, and are most pronounced during sympathetic activation. It is likely that the haemodynamic changes during physical or psychological challenge will be most strongly influenced by rebound phenomena within the first two or three days following withdrawal. In many research protocols, it is not appropriate, feasible or ethical to stop antihypertensive therapy and wait for three to four weeks before examination. This leads to the conclusion that medicated hypertensives should be tested while undergoing chronic pharmacological therapy. This strategy, however, may cause problems in the analysis of haemodynamic changes during stress testing. Different therapeutic regimens vary in their impact on cardiovascular reactivity and on behaviour (Schmieder et al 1983).

CONCLUSIONS

When studies are conducted comparing the stress responses of hypertensives and normotensives, it is most appropriate to examine diagnosed hypertensives who have not undergone any drug therapy. On the other hand, if hypertensives with a long history of disease are studied, stress testing should not be performed earlier than four weeks after withdrawal of antihypertensive medications, or else they should be tested during chronic medication. These options seem more satisfactory than testing patients shortly after drug withdrawal, when rebound phenomena will compromise responses.

REFERENCES

Avery GS (1980) Drug Treatment. Principles and Practice of Clinical Pharmacology and Therapeutics. ADIS Press, Sydney, p1295

Bengtsson C, Larsson Y, Panfilov V, Sivertsson R, Aström H (1981) Haemodynamic effect of withdrawal of long-term treatment with ß-adrenoceptor blocking agents in subjects with essential hypertension. Clin Sci 61:421S-423S

Bolli P, Bühler FR, Raeder EA, Amann FW, Meier M, Rogg H, Burckhardt D (1981) Lack of ß-adrenoceptor hypersensitivity after abrupt withdrawal of long-term therapy with oxprenolol. Circulation 64:1130-1134

Boudoulas H, Lewis RP, Kates RE, Dalamangas G (1977) Hypersensitivity to adrenergic stimulation after propranolol withdrawal in normal subjects. Ann Intern Med 87:433-436

Garrett BN, Kaplan NM (1980) Plasma renin activity suppression. Duration after withdrawal from ß-adrenergic blockade. Arch Intern Med 140:1316-1318

Garbus SB, Weber MA, Priest RT, Brewer DD, Hubbell FH (1979) The abrupt discontinuation of antihypertensive treatment. J Clin Pharmacol 19:476-486

Gerber FG, Nies AS (1979) Abrupt withdrawal of cardiovascular drugs (editorial). New Engl J Med 301:1234-1235

Geyskes GG, Boer P, Dorhout Mees EJ (1979) Clonidine withdrawal. Mechanism and frequency of rebound hypertension. Brit J Clin Pharmacol 7:55-62

Gilfrich HJ, Rahn KH (1981) Absetzphänomene bei Beendigung einer Therapie mit ß-Rezeptorenblockern. Med Welt 32:1036-1037

Goldberg AD, Raftery EB, Wilkinson P (1977) Blood pressure and heart rate and withdrawal of antihypertensive drugs. Brit Med J 1:1243-1246

Hjemdahl P, Olsson G (1982) Rebound phenomena following withdrawal of long-term ß-adrenoceptor blockade. Acta Med Scand (Suppl) 665:43-47

Krämer B, Kübler W (1981) Absetzphänomene nach chronischer ß-Blockade. Dtsch Med Wochenschr 109:465-471

Kristensen BO, Steiness E, Weeke J (1978) Propranolol withdrawal and thyroid hormones in patients with essential hypertension. Clin Pharmacol Ther 23:624-629

Maling TJ, Dollery CT (1979) Changes in blood pressure, heart rate and plasma noradrenaline concentration after sudden withdrawal of propranolol. Brit Med J 2:366-367

Maslowski AH, Nicholls MG, Ikram H, Espinger EA, Turner JG (1981) Haemodynamic, hormonal and electrolyte responses to withdrawal of long-term captopril treatment for heart failure. Lancet II: 959-961

Petersen OL, Mikkelsen E, Nielsen JL, Christensen NJ (1979) Abrupt withdrawal of ß-blocking agents in patients with arterial hypertension. Effect on blood pressure, heart rate and plasma catecholamine and prolactin. Europ J Clin Pharmacol 15:215-217

Ragno RE, Langlois S (1982) Comparison of withdrawal phenomena after propranolol, metaprolol and pindolol. Brit J Clin Pharmacol 13:345s-351s

Schmieder R, Friedrich G, Neus H, Rüddel H, von Eiff AW (1983) The influence of ß-blockers on cardiovascular reactivity and Type A behavior pattern in hypertensives. Psychosom Med 45:417-423

Shand DG (1975) Propranolol withdrawal (editorial). New Engl J Med 293:449-450

Whitsett TL (1983) Abrupt cessation of treatment with centrally acting antihypertensive agents. A review. Chest 2 (suppl): S400-S402

CHAPTER 12

Methodology of Studying the Catecholamine Response to Stress

Marcia M Ward and Ivan N Mefford

Department of Behavioral Medicine, SRI International, 333 Ravenswood Avenue, Menlo Park, California, 94025 USA
Chemistry Department, Boston College, Chestnut Hill, Massachusetts 02167, USA

SUMMARY

The catecholamines (epinephrine, norepinephrine and dopamine), as measured in urine or plasma, increase dramatically in response to a variety of stressors including emotional stimuli, cognitive challenges, and physical activity. Measuring the catecholamine response to stress is often desirable because it provides the most direct assessment of sympathetic nervous system activity. This chapter discusses the methodology for measuring catecholamine responses to stress. Because catecholamines are so responsive to a variety of stimuli, a number of factors which must be carefully controlled in studies of catecholamine responses are discussed. The techniques for collecting both urine and plasma samples are presented along with a discussion of the appropriate timing of samples. Finally, procedures for preparing and storing urine and plasma samples and for assaying catecholamine levels are presented.

INTRODUCTION

The catecholamines (epinephrine or adrenaline, norepinephrine or noradrenaline, and dopamine) are very responsive to stress. They increase dramatically to a variety of stressors including emotional stimuli, cognitive challenges and physical activity. Catecholamines are released peripherally from the adrenal medulla and nerve terminals as an important link in the activation chain of the sympathetic nervous system. Catecholamines released from the adrenal medulla (epinephrine and norepinephrine) immediately enter the blood stream and act as circulating hormones; those released from nerve terminals (norepinephrine and dopamine) act as neurotransmitters and their overflow makes its way into the circulation. Circulating catecholamines are absorbed by the liver, metabolised, and excreted in urine.

This chapter discusses the methodology for assessing catecholamine responses to stress. First, factors that affect catecholamine levels and which must be controlled in studies of responses to stress are discussed. Then, the techniques for collecting both urine and plasma samples are presented along with a discussion of the appropriate timing of samples. Finally, procedures for assaying catecholamine levels are described.

FACTORS AFFECTING CATECHOLAMINE LEVELS

Because catecholamines are so responsive to stress and other stimuli that affect sympathetic nervous system activation, these factors must be carefully controlled. This section describes (1) the individual factors that must be controlled when comparing subjects; (2) the factors that have a chronic effect on catecholamines and must be considered when an experimental session is scheduled; (3) the long-term factors to be controlled before sample collection; and (4) the acute factors to be controlled during sample collection

Individual Factors to be Controlled when Comparing Subjects

There are several individual factors which have been found to correlate with catecholamine levels. These include the age and sex of the subject, and a large number of disease states. These factors must be controlled when comparing subjects or groups.

Age: Age is positively correlated with resting norepinephrine levels, and the norepinephrine response to exercise and mental stress appears to be larger in older subjects (Barnes et al 1982; Lake et al 1976). Epinephrine resting levels and response to mental stress, however, do not appear to change with age (Barnes et al 1982).

Sex: Sex differences in catecholamine levels during rest are slight. However, during most laboratory and real-life stressors, males secrete significantly more catecholamines, especially epinephrine, than do females even when performance is identical. This pattern remains, but is diminished, when 'nontraditional' females such as female engineers, lawyers, or bus drivers are compared to their male counterparts (Frankenhaeuser 1983).

Cardiovascular disease states: These significantly alter catecholamine levels. Catecholamines are significantly elevated during and after myocardial infarction and in patients with congestive heart failure. There is a large literature indicating elevated catecholamines in hypertensives, but this conclusion remains controversial (ie Goldstein 1981). Catecholamines may be consistently elevated only for a small percentage of young established hypertensives who have a defect in neuronal uptake of epinephrine (Esler et al 1981).

Psychiatric disease states: These are hypothesised to be related to defects in the catecholamine system. Individuals with panic disorder, anxiety, depression and schizophrenia may have altered catecholamine levels.

Chronic Factors that Affect Testing Schedules

There are several factors which have a chronic effect on catecholamine levels that must be considered when scheduling subjects for testing. These include daily cycles and monthly cycles. It is also possible that there is a seasonal variation in catecholamine levels.

Circadian rhythm: Catecholamines appear to be governed by a circadian rhythm. Both epinephrine and norepinephrine have been shown to be lowest in the early morning hours during sleep and to be highest during the daytime (Turton and Deegan 1974).

Menstrual cycle: The menstrual cycle may affect catecholamine levels in females. Norepinephrine has been found to show a sharp, transient peak one to three days preceding ovulation and again one ray after ovulation (Badano et al 1978). However, others have found no relationship between plasma norepinephrine levels and the time of the menstrual cycle (Zacur et al 1978).

These factors having a chronic effect can best be controlled by picking one time of testing for all subjects. It is usually recommended to test all subjects early in the workday when individual differences in activity levels are likely to be minimised. Females should probably be tested during the follicular or luteal phases of their menstrual cycle.

Factors to be Controlled before Sample Collection

There are a number of factors that dramatically affect catecholamine release for several hours or days. These factors can produce an increase or decrease in the basal level of catecholamines. They can also change the catecholamine response to other stimuli. Since these factors have an extended period of action they must be carefully monitored during any study of catecholamines and for several hours or even days prior to sample collection.

Caffeine: Caffeine has been reported to increase plasma epinephrine by 207% and plasma norepinephrine by 75% (Robertson et al 1978). Both catecholamines remain elevated for at least three hours after ingestion of caffeine.

Alcohol: Alcohol has been found to produce elevated epinephrine, norepinephrine and dopamine levels in urine and to increase catecholamine synthesis (Anton 1965; Gitlow et al 1976).

Sodium: Sodium intake affects catecholamines in both normotensives and hypertensives. Both low and high salt intake have been related to elevated plasma epinephrine and norepinephrine levels in normotensives, with low sodium producing the greatest elevations (Nicholls et al 1980; Romoff et al 1978).

Nicotine: Smokers have higher epinephrine excretion than nonsmokers but no difference in norepinephrine levels (Westfall and Watts 1964). In one well-controlled study (Cryer et al 1976) plasma epinephrine was shown to peak at ten minutes after the start of smoking and then to decline gradually but remain elevated for at least 30 minutes; plasma norepinephrine was shown to peak at 15 minutes after the start of smoking but then to fall rapidly to within normal limits by 20 minutes.

Medication: Numerous drugs affect catecholamine levels. The most important of these are ß-blockers and medications containing epinephrine which both alter levels dramatically. Other medications that may affect catecholamines are diuretics, antipsychotics, L-DOPA, antihistamines, antimanics, amphetamines, antidepressants, and anxiolytics (Ciaranello and Patrick 1977).

Exercise: Exercise dramatically increases epinephrine, norepinephrine and dopamine, usually in direct relationship to the amount of isometric and aerobic activity (see Ward et al 1984). Catecholamine levels can continue to increase further after aerobic exercise has been stopped (Dimsdale et al 1984).

The factors listed above all affect catecholamines for some period of time, usually hours. Subjects should be asked about their daily exercise activity and intake of medication, alcohol, caffeine, sodium, and nicotine. Usually abstinence from these substances for eight or more hours is recommended. Subjects find abstinence easiest if they are tested in the morning so that their usual schedule of intake is minimally disrupted. However, some smokers find abstinence from nicotine for even a few hours very difficult. Unfortunately the effect of nicotine withdrawal on catecholamines is not known. In addition, it is recommended that subjects not be tested in a fasting state since hunger may increase catecholamines (Lake et al 1976).

Acute Factors to be Controlled During Sample Selection

There are a number of factors that acutely affect catecholamines which must be controlled during the experiment. These factors include movement and posture changes, food and drink intake, and emotional stimuli. The procedure for controlling these factors depends somewhat on whether the experiment involves plasma or urine samples.

Movement and posture changes: These dramatically affect epinephrine, norepinephrine and dopamine. Rising from a supine to a standing position causes plasma catecholamines to double in value (see Ward et al 1984 for review). A sitting posture produces higher levels than does a supine posture (Mefford et al 1981). During studies where plasma samples are drawn, posture can and should be carefully monitored. During studies where urine samples are collected, posture and movement should be recorded in a diary. If group differences on urine catecholamine levels are of interest, then the subjects in each group should be comparable in the type and amount of exercise that they usually engage in and the movement that they are likely to experience.

Food and drink intake: Foods and drinks containing caffeine, alcohol, and sodium produce elevations in catecholamines, as indicated above. Volume expansion and contraction significantly affect catecholamines in normotensives (Henry et al 1980). Food and drinks are not usually consumed during plasma collection, but are often consumed during studies involving urine collection. in such studies everything consumed and the quantity should be recorded in a diary.

Emotional stimuli: These have been shown to produce elevations in both epinephrine and norepinephrine (see Frankenhaeuser 1975). During plasma sampling the experimental environment should be quiet and the subject should be restricted from extraneous stimuli that are not under direct control. During studies where urine samples are collected, significant emotions (fear, anxiety, anger, etc) should be recorded in a diary.

CHOICE OF SAMPLE COLLECTION PROCEDURE AND TECHNIQUES

Catecholamine responses to stressors can be measured in either urine or plasma; however, the values obtained are not necessarily comparable. Januszewicz et al (1979) found that plasma and urine norepinephrine levels were significantly correlated, but plasma and urine epinephrine values were unrelated. However, Akerstedt et al

(1983) found stronger correlations for epinephrine than for norepinephrine. There are several reasons why plasma and urine catecholamine values can differ dramatically. Because the half-life of catecholamines is a minute or two (Fitzgerald et al 1979; La Brosse et al 1961; Silverberg et al 1978), levels measured in plasma represent activity in the preceding few minutes. Urine, on the other hand, represents a pooling of catecholamine levels over the entire period of time since the bladder was last emptied - usually several hours. Thus if catecholamine responses to short-term stressors are desired plasma samples are essential. If catecholamine responses to long-term, chronic stressors are of interest, then urine samples are preferable.

In the following section the advantages and disadvantages and techniques for collecting plasma and urine samples are presented. While the ease of technique to be employed is often an important factor, ultimately the research question to be answered dictates which collection procedure and sampling rate is used. Issues of sampling rate are discussed in a later section.

Urine Samples - Advantages and Disadvantages

Measuring catecholamine levels in urine offers several advantages over plasma. The procedure is usually very convenient for the experimenter and presents no risk to the subject. It is the procedure of choice when levels are desired over an extended period to study the effects of chronic stressors. Catecholamine levels can be measured for hours or even days, or as frequently as every few hours if fluid intake is high.

Measuring catecholamine levels in urine also has several disadvantages. The subject needs to carry around a urine collection bottle which can be embarrassing to some individuals. Proper sample collection takes some care on the part of the subject, so instruction on the proper collection procedure is recommended. In general, the collection procedure is out of the experimenter's hands which means that extraneous factors such as movement, food and drink intake cannot be monitored, and compliance with collection procedures is unverifiable. A further disadvantage is that the catecholamines which appear in the urine were released some time before, and the specific time of the catecholamine release cannot be determined.

Plasma Samples - Advantages and Disadvantages

Measuring catecholamines in plasma, like urine, has its advantages and disadvantages. This is the procedure of choice when levels are desired at specific moments to study the effects of acute stressors. Blood samples can be drawn as frequently as every few minutes. Catecholamine levels found in plasma represent activity in the minutes preceding the sample collection - so the catecholamine level can be matched to specific activity. Also, because the subject is watched during plasma collection, intervening variables can be carefully monitored.

The disadvantages of measuring catecholamines in plasma samples include the need for medical personnel (phlebotomist, nurse or physician) to perform the venipuncture. There is some minimal risk to the subject in collecting blood samples. Additionally, venipuncture causes pain in most subjects and has been shown to produce elevated catecholamines (Carruthers et al 1970; Lake et al 1976).

Urine Sample Collection Methodology

Urine samples are usually collected and pooled over an eight-hour, 12-hour or 24-hour period. When pooling is done over a long period, a large quantity of urine is collected. The subject must be provided with suitable containers to make both urine collection and storage easy. If the collection period is long, then several collection bottles must be provided. An alternative to using large collection bottles is a simple portable semi-automatic urine sampling cup with divided partitions (Tochikubo et al 1983). This automatically separates out a small subsample of each urine sample for storage and allows the excess to be discarded. This device and the subsample of urine is carried in a vinyl case, minimising the potential for embarrassment of the subject.

When urine samples are to be collected over an extended period of time, the subject must be instructed to collect urine from each urination, to comply with all instructions, and to document certain events in a diary. The subject should record all exercise or excessive movement, everything consumed, and any events which evoked an emotional response.

When several urine samples are required in a given time period, or when urine samples are desired that represent a relatively short time period, then the production of urine must be maximised. This can be accomplished by having the subject drink fluids prior to and during the collection period and by maintaining a low sodium diet. However, these manoeuvres can affect catecholamine levels (Henry et al 1980) so volume expansion and sodium intake must be monitored and kept comparable across test groups or conditions.

Plasma Sample Collection Methodology

Measuring catecholamines in plasma necessitates drawing blood which involves venipuncture. Venipuncture causes pain in most subjects, and hence produces elevated catecholamines (Carruthers et al 1970; Lake et al 1976). The dramatic rise in catecholamines to venipuncture dictates that single or repeated venipunctures should never be employed. Instead, an indwelling catheter or butterfly needle with a 'heparin lock' should be inserted at least 20 minutes before blood is drawn for determination of baseline values.

The subject may be tested in a supine or sitting position. Any movement from this position must be recorded as it will affect catecholamine levels.

Many different types of stressors can be easily administered. If more than one stressor is used in a test session, then the rest period between stressors must be of adequate duration to permit full recovery. The amount of time needed for catecholamine levels to return to baseline depends on the type of stressor used. As shown in Fig 1, recovery after a purely mental stressor may occur in ten minutes, whereas the recovery following cold pressor may require 20 minutes. Recovery following strenuous aerobic exercise can be quite lengthy especially for unconditioned subjects. Also the recovery for epinephrine is faster than for norepinephrine (Ward et al 1983).

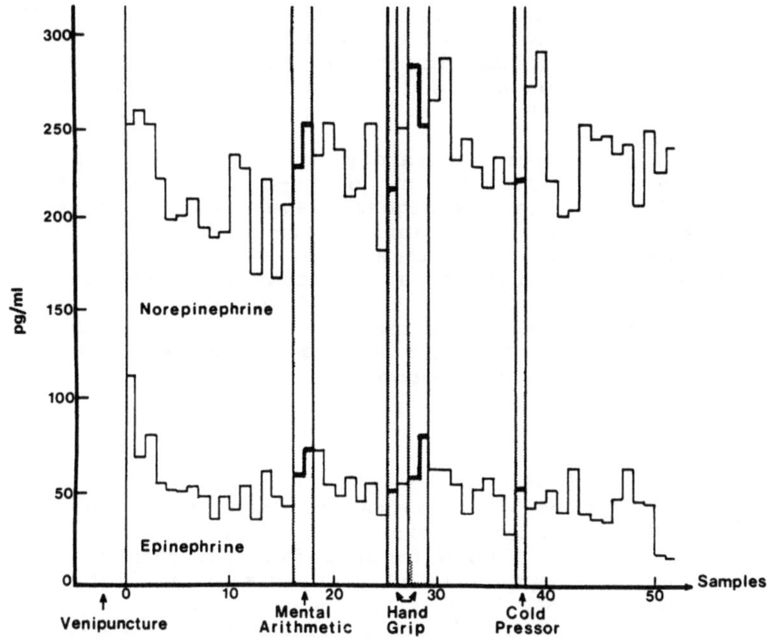

Fig 1. Median levels of norepinephrine (top line) and epinephrine (bottom line) across 53 two-minute samples of continuously collected blood. The cross-hatched areas indicate samples during mental arithmetic, handgrip, and cold pressor.

If several stressors are administered in one test session, then a baseline sample should be collected at the conclusion of each recovery period, prior to the next stressor. This is important because circadian rhythm or the accumulated effect of incomplete recovery from multiple stressors may cause a significant trend in resting levels. Using repeated pre-task baselines prevents a stressor from being compared to a baseline that occurred some considerable time before which is no longer comparable to resting levels. It can also preclude the need for using multiple presentation orders of stressors.

TIMING OF PLASMA SAMPLES

Timing of plasma samples is a crucial variable. The half-life of epinephrine in humans is approximately 1.2 minutes (La Brosse et al 1961) and the half-life of norepinephrine is approximately 2.5 minutes (Fitzgerald et al 1979; Silverberg et al 1978). The catecholamines are thought to be released intermittently; consequently, the appearance of norepinephrine and, particularly, epinephrine in blood appears to be rapid and transitory.

If a chosen stressor is more than a few minutes long, then several samples should be collected during the stressor. The plasma epinephrine level at the beginning of public speaking, for example, is significantly higher than that obtained during the middle phase of a speech (Dimsdale and Moss 1980). On the other hand, plasma epinephrine and norepinephrine levels continue to rise after aerobic exercise is stopped (Dimsdale et al 1984). The specific point during each stressor when blood is sampled must be given careful consideration.

Data from several studies in our laboratory on continuous sampling of catecholamines during a variety of short-duration stressors suggest that epinephrine reaches its highest level within the first two minutes of a stressor while norepinephrine reaches its peak level six to eight minutes later, in this case, after the stressor was completed. These data (results from one study are shown in Fig 1) suggest that it might be preferable to draw two blood samples - one within the first two minutes of a stressor for analysis of the peak epinephrine response, and one four to eight minutes later for analysis of the peak norepinephrine response.

CONTINUOUS PLASMA WITHDRAWAL

One solution to the problem of when to draw a blood sample for the most sensitive analysis of catecholamine levels is to draw blood continuously. This is accomplished with an indwelling catheter and an effusion pump. One such pump that is commercially available is produced by Cormed (Medina, New York). It is quite small (10x9cm) and easily portable so that it can be used for ambulatory purposes. Cormed also supplies sterile, antithrombogenic-treated butterfly needles or catheters for use with their pump. This pump has been most frequently used for collecting blood for catecholamine analysis by Dimsdale and his associates (Dimsdale and Moss 1980; Dimsdale et al 1984). Two disadvantages to the Cormed pump are that: (1) the collection tube must be changed at the conclusion of each sample period; and (2) the flow rate is predetermined with only two speeds available.

Alternatives to the effusion pump produced by Cormed are available. None are small enough to be used for ambulatory purposes, but they can be easily used in the laboratory setting. One such system includes a Gilson FC-100 peristaltic pump as described by Ward et al (1983). For hands-off sample acquisition, the pump can be connected to a microfractionator which automatically pools the blood into samples according to a set time or drip rate. A wide range of tubing sizes, pump speeds, and microfractionator collection rates makes this system very flexible.

Continuous blood withdrawal permits all of the catecholamine activity to be monitored (an example of this type of data is shown in Fig 1), and is thus the only means to assure that no important catecholamine responses are missed. However, this procedure is encumbered with difficulties. The most consistent problem arises from the physiological barriers to continuous blood withdrawal. Continuous blood withdrawal is prevented by platelet formation, collapsing of the vein, and vasoconstriction during stressors. The chances of successful blood withdrawal are enhanced in individuals who have large veins, a good history with blood drawing, and a calm attitude about the procedure. Even in such individuals, blood withdrawal may not be initiated or the flow may diminish and stop any time. Unfortunately, manipulating the catheter can be stressful for the subject and produce elevations in catecholamine levels, so such manipulation must be kept to a minimum. Of course, an additional drawback with this technique is that the large number of samples that are produced make chemical analysis a costly endeavour. However, certain research questions on the speed of catecholamine response and recovery rate warrant using this technique.

SAMPLE PREPARATION AND STORAGE

Sample handling prior to isolation of catecholamines is fairly straightforward. While it has been recommended that blood be kept chilled and rapidly spun, separated, and frozen (Lake et al 1976), others have shown that catecholamines are stable in freshly collected blood or plasma until frozen (Falconer et al 1982). However, antioxidants and anticoagulants are generally included in the collection tube. Ethylenediaminetetragetic acid (EDTA) has both antithrombogenic effects and also complexes metal ions which may be present in blood, eliminating the metal ion catalysed oxidation of catecholamines. As a further precaution, glutathione is frequently included as an antioxidant.

Storage of plasma samples is usually accomplished by freezing at -80°C. When stored at -80°C, catecholamines are stable indefinitely and large numbers of samples may be collected over a long time period. Storage at normal freezer (-10°C) temperatures is not recommended if analysis cannot be accomplished within a few days. Multiple freezing and thawing is also not recommended unless a suitable internal standard such as dihydroxybenzylamine has been added which can be used to compensate for loss through oxidation.

Urinary catecholamines are also stored at -80°C if long storage times are necessary. Acidification to approximately pH 3.0 and addition of EDTA and an antioxidant (ascorbic acid, glutathione or sodium metabisulfite) are recommended as precautions. Again, samples may be safely stored for long periods of time, perhaps years, if suitable precautions against oxidation are taken and the samples are maintained at -80°C.

METHODOLOGY OF SAMPLE ASSAYING PROCEDURES

A variety of analytical techniques have been employed over the years for the quantitative determination of catecholamines in urine and plasma. A brief review of these techniques and their advantages and disadvantages are discussed.

Bioassay Procedures

Earliest methods relied upon bioassay procedures. These techniques depended upon the action of the amine on muscle contraction or another physiological parameter as an identifying mechanism and the extent of this action as a measure of the quantity of the amine present. Bioassay techniques for biogenic amines are quite sensitive but generally lack specificity and the reproductibility of the biological preparation is poor.

Fluorescence

Several spectroscopic methods have been used for catecholamine determination, the most pervasive of which has been the trihydroxyindole (THI) fluorescence technique (Lund 1949; Weil-Melherbe and Bigelow 1968). The THI method has been frequently used for urine samples but lacks the inherent sensitivity to allow detection of catecholamines in small volumes of blood plasma. Further complications of this technique are the similarities of the absorption and emission spectra of epinephrine and norepinephrine.

Radioenzymatic Techniques

Radioenzymatic techniques have found widespread application over the past decade. For the determination of catecholamines, these methods are generally based on the incubation of a plasma extract with either an O-methyl or N-methyl transferase enzyme and the methyl donor, S-adrenosyl methionine (SAM). By using SAM labelled with [^{3}H] methyl of high specific activity, one obtains selective formation of the radioactive N or O-methylated catecholamine derivatives. The radioactive products are then isolated by selective extraction or chromatographic techniques and quantitated by measuring the radioactivity. A number of modifications of this general procedure are available. Most workers find detection limits in the low picogram range (Peuler and Johnson 1977; Saito et al 1983). Radioenzymatic techniques are widely used for routine determination of catecholamines and metabolites and represent one of the more popular methods.

Combined Gas Chromatography-Mass Spectrometry (GC-MS)

This is a powerful technique as it can provide chemical identification of the species of interest. Catecholamines are derivatised to provide stability and volatility. The derivatised species are separated using gas chromatography and the effluent from the GC column is then passed into the source of the mass spectrometer where the molecules are ionised. The resulting ions are collimated and passed into the mass analyser where they are separated based on their mass-to-charge ratio. A spectrum of ions of varying intensities is obtained which can be used to give unequivocal identification. This general approach has been used by several authors for analysis of plasma catecholamines (Karoum 1983; Wang et al 1975). Combined gas chromatography-mass spectrometry requires relatively expensive and complex instrumentation and a dedicated operator. Extraction and derivatisation techniques are also fairly complex, yet only this technique offers unequivocal identification.

High Performance Liquid Chromatography with Electrochemical Detection (LCEC)

With LCEC, catecholamines are selectively extracted from plasma with aluminium oxide (Anton and Sayre 1962) or by a borate complexation procedure (Smedes et al 1982). Once extracted, the catecholamines are separated using either cation exchange or reverse phase liquid chromatography and detected electrochemically by oxidation at a graphite electrode (Mefford et al 1981). Each step insures selectivity, the extraction being selective for catechols, the chromatographic separation selective for amines and the low oxidation potential selective for the benzene ring substituted with two or more electron donating groups. Recent advances in chromatographic procedures have considerably enhanced the limits of detection for LCEC (Caliguri and Mefford 1984). Limits of detection of 0.1 to 0.5 picograms for catecholamines have been reported. Additionally, high speed liquid chromatographic techniques can decrease analysis times to less than five minutes per sample for all three catecholamines.

All of the latter three techniques have been used routinely and successfully for measurement of catecholamines in small plasma samples (less than 2ml). Limits of detection are comparable although radioenzymatic and LCEC approaches are considerably less expensive and tedious than the GC-MS procedures.

CONCLUSIONS

Measuring the catecholamine response to stress provides the most direct assessment of sympathetic nervous system activity. The methodology for measuring catecholamines is technically straightforward and in urine is noninvasive. Catecholamines are of considerable interest because of their direct role in the development of atherogenesis and their link to coronary heart disease and hypertension. Measuring catecholamines adds some extra effort and expense to a research protocol, but the payoff in direct information about sympathetic nervous system activation often outweighs the added technical effort.

REFERENCES

Akerstedt T, Gillberg M, Hjemdahl P, Sigurdson K, Gustavsson I, Daleskog M, Pollare T (1983) Comparison of urinary and plasma catecholamine responses to mental stress. Acta Physiol Scand 117: 19-26

Anton AH (1965) Ethanol and urinary catecholamines in man. Clin Pharmacol Ther 6: 462-469

Anton AH, Sayre DF (1962) A study of the factors affecting the aluminum-oxide-trihydroxide procedure for the analysis of catecholamines. J Pharm Exp Ther 138: 360-374

Badano AR, Nagle CA, Casas PRF, Miechi H, Mirkin A, Turner DE, Aparicio N, Rosner JM (1978) Plasma levels of norepinephrine during the preovulatory period in normal women. Amer J Obstet Gynecol 131: 299-303

Barnes RF, Rasking M, Gumbrecht G, Halter JB (1982) The effect of age on the plasma catecholamine response to mental stress in man. J Clin Endocrin Metabl 24(1): 64-69

Caliguri EJ, Mefford IN (1984) Femtogram detection limits for biogenic amines using microbore HPLC with electrochemical detection. Brain Res 296: 156-159

Carruthers M, Taggart P, Conway N, Bates D, Somerville W (1970) Validity of plasma catecholamine estimations. Lancet 1: 62-67

Ciaranello RD, Patrick RL (1977) Catecholamine neuroregulators. In: Barchas JD, Berger PA, Ciaranello RD, Elliott GR (eds) Psychopharmacology, from Theory to Practice. Oxford University Press, New York, p16

Cryer PE, Haymond MW, Santiago JV, Shah SD (1976) Norepinephrine and epinephrine release and adrenergic mediation of smoking-associated hemodynamic and metabolic events. New Eng J Med 295: 573-577

Dimsdale JE, Hartley LH, Guiney T, Ruskin JN, Greenblatt D (1984) Postexercise peril: plasma catecholamines and exercise. J Amer Med Assoc 251(1): 630-632

Dimsdale JE, Moss J (1980) Short-term catecholamine response to psychological stress. Psychosom Med 42(5): 493-497
Esler M, Jackman G, Bobik A, Leonard P, Kelleher D, Skews H, Jennings G, Korner P (1981) Norepinephrine kinetics in essential hypertension. Hypertension 3: 149-156
Falconer AD, Lake D, Macdonald IA (1982) The measurement of plasma noradrenaline by high-performance liquid chromatography with electrochemical detection: an assessment of sample stability and assay reproducibility. J Neurosci Methods 6: 261-271
Fitzgerald GA, Hossmann V, Hamilton CA, Reid JL, Davies DS, Dollery CT (1979) Interindividual variation in kinetics of infused epinephrine. Clin Pharm & Ther 26(6): 669-675
Frankenhaeuser M (1975) Experimental approaches to the study of catecholamines and emotion. In: Levi L (ed) Emotions - Their Parameters and Measurement. Raven Press, New York, p209
Frankenhaeuser M (1983) The sympathetic-adrenal and pituitary-adrenal response to challenge: comparison between the sexes. In: Dembroski TM, Schmidt TH, Blumchen G (eds) Biobehavioral Bases of Coronary Heart Disease. Karger, Basel, p91
Gitlow SE, Dziedzic LM, Dziedzic SW, Wong BL (1976) Influence of ethanol on human catecholamine metabolism. Ann NY Acad Sci 273: 263-279
Goldstein DS (1981) Plasma norepinephrine in essential hypertension. Hypertension 3: 48-52
Henry DC, Luft FC, Weinberger MH, Fineberg NS, Grim CE (1980) Norepinephrine in urine and plasma following provocative maneuvers in normal and hypertensive subjects. Hypertension 2: 20-28
Januszewicz W, Sznajderman M, Wocial B, Feltynowski T, Klonowicz T (1979) The effect of mental stress on catecholamines, their metabolites and plasma renin activity in patients with essential hypertension and in healthy subjects. Clin Sci 57: 229s-231s
Karoum F (1983) Mass fragmentography in the analysis of biogenic amines: a clinical, physiological and pharmacological evaluation. In: Parvez S, Ngatsu T, Nagatsu I, Parvez H (eds) Methods in Biogenic Amine Research. Elsevier, Amsterdam, p235
La Brosse EH, Mann JD, Kety SS (1961) The physiological and psychological effects of intravenously administered epinephrine and its metabolism, in normal and schizophrenic men. J Psychiat Res 1: 68-75
Lake CR, Ziegler MG, Kopin IJ (1976) Use of plasma norepinephrine for evaluation of sympathetic neuronal function in men. Life Sciences 18: 1315-1326
Lund A (1949) Fluorometric determination of adrenaline in blood: new sensitive and specific method. Acta Pharmacol et Toxicol 5: 231-247
Mefford IN, Ward MM, Miles L, Taylor B, Chesney MA, Keegan DL, Barchas JD (1981) Determination of plasma catecholamines and free 3,4 dihydroxyphenylacetic acid in continuously collected human plasma by high performance liquid chromatography with electrochemical detection. Life Sciences 28: 477-483
Nicholls MG, Kiowski W, Zweifler AJ, Julius S, Schork MA, Greenhouse J (1980) Plasma norepinephrine variations with dietary sodium intake. Hypertension 2: 29-32
Peuler JD, Johnson GA (1977) Simultaneous single isotope radio-enzymatic assay of plasma norepinephrine, epinephrine and dopamine. Life Sciences 21: 625-636
Robertson D, Frolich JC, Carr RK, Watson JT, Hollifield JW, Shand DG, Oates JA (1978) Effect of caffeine on plasma renin activity, catecholamines, and blood pressure. New Engl J Med 298: 181-186

Romoff MS, Keusch G, Campese VM, Wang M, Friedler RM, Weidmann P, Massey SG (1978) Effect of sodium intake on plasma catecholamines in normal subjects. J Clin Endocrinol Metab 48: 26-31

Saito H, Minami M, Togashi H, Koite Y (1983) Radioenzymatic assays for catecholamine determination. In: Parvez S, Nagatsu T, Nagatsu I, Parvez H (eds) Methods in Biogenic Amine Research. Elsevier, Amsterdam, p285

Silverberg AB, Shah SO, Haymond MW, Cryer PE (1978) Norepinephrine: hormone and neurotransmitter in man. Am J Physiol 234(3): E252-E256

Smedes F, Kraak JC, Poppe J (1982) Simple and fast solvent extraction system for selective and quantitative isolation of adrenaline, noradrenaline and dopamine from plasma and urine. J Chromatography 231: 25-39

Tochikubo O, Uneda S, Kaneko Y (1983) Simple portable device for sampling a whole day's urine and its application to hypertensive outpatients. Hypertension 5: 270-274

Turton ME, Deegan T (1974) Circadian variations of plasma catecholamine, cortisol and immunoreactive insulin concentrations in supine subjects. Clin Chim Acta 55: 389-397

Wang MT, Imai K, Yoshioka M, Tamura E (1975) Gas-liquid chromatographic and mass fragmentographic determination of catecholamines in human plasma. Clin Chim Acta 63: 13-19

Ward MM, Mefford IN, Black GW, Bortz WB (1984) Exercise and plasma catecholamine release. In: Fotherby K, Pal SB (eds) Exercise Endocrinology. Walter de Gruyter & Co, Berlin

Ward MM, Mefford IN, Parker SD, Chesney MA, Taylor CB, Keegan DL, Barchas JD (1983) Epinephrine and norepinephrine responses in continuously collected human plasma to a series of stressors. Psychosom Med 45(6): 471-486

Weil-Malherbe H, Bigelow LB (1968) The fluorometric estimation of epinephrine and norepinephrine. An improved modification of the trihydroxindole method. Analyt Biochem 22: 321-334

Westfall TC, Watts DT (1964) Catecholamine excretion in smokers and nonsmokers. J Appl Physiol 19(1): 40-42

Zacur HA, Tyson JE, Ziegler MG, Lake CR (1978) Plasma dopamine -B-hydroxylase activity and norepinephrine levels during the human menstrual cycle. Amer J Obstet Gynecol 130: 148-151

CHAPTER 13

Phasic Heart Rate and Vagal Reactivity: Methods and Interpretations

Riek Somsen

Subfaculteit Psychologie, Vrije Universiteit, De Boelelaan 1115, NL-1081 HV AMSTERDAM, The Netherlands

SUMMARY

Phasic heart rate changes have been mainly studied in relation to attentional and task-related information processing in non-stressful experimental situations. Most studies of individual differences in autonomic activity in response to stress situations have not measured phasic heart rate responses; they have been predominantly concerned with effects of induced stress on heart rate level and other cardiovascular sympathetic indices. The present paper will discuss some methodological aspects of the measurement of phasic heart rate, some interpretations of characteristic response types, the relative contribution of changes in parasympathetic and sympathetic activity and other autonomic changes, and the potential relevance of the phasic heart rate response for studies of individual differences in cardiovascular activity during stress.

ASSESSMENT OF PHASIC HEART RATE RESPONSES

Investigators of phasic heart rate, that is beat-by-beat or second-by-second variations in heart rate, are in general interested in the short-term effects of temporal stimuli or responses on cardiac activity. These are usually assessed in highly controlled experimental situations. Phasic changes in beat frequency are typically time locked to some event that has special meaning for the subject. These changes are not necessarily elicited by external stimulation, since they may also occur in anticipation of such stimulation.

Figure 1 shows beat-by-beat changes in cardiac period and second-by-second changes in cardiac rate recorded in an unsignalled choice reaction time task, averaged over about 120 trials in 15 subjects. The two representations are based on the same data set. Cardiac deceleration, ie, increase in cardiac period (left panel) and decrease in cardiac rate (right panel), occurred in anticipation of the imperative tone stimulus and acceleration occurred after stimulus presentation and the subsequent execution of the rapid motor response.

Obviously, the beat-by-beat measure has no fixed time axis across successive trials and subjects. A smaller real time sample is obtained from trials and subjects with rapid heart rates than from trials and subjects with low heart rate levels. The measure implicitly assumes that an event-related cardiac response follows an organismic cardiac clock. In contrast, expression in real time

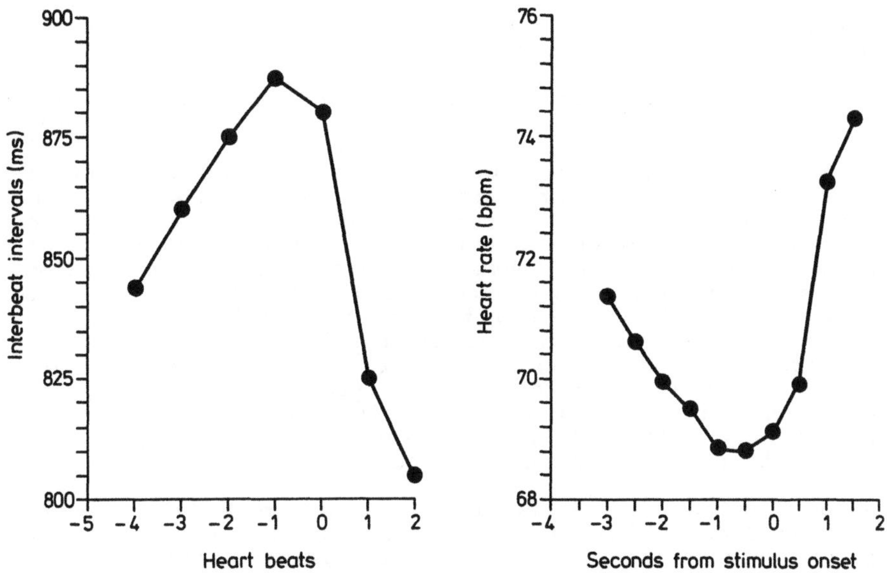

Fig 1. Averaged phasic heart rate response in an unsignalled choice reaction time task represented either in interbeat interval (msec) and successive beats (left panel) or in heart rate (beats per minute) and half seconds (right panel). Four beat intervals before (-), one at the time of (0), and two beat intervals after the onset of the stimulus (+) are depicted on the left panel. The right panel shows the heart rate responses from three seconds before until two seconds after the time of stimulus onset (time 0).

units assumes time dependency of cardiac responses. This latter sampling technique is more in accordance with centrally mediated attentional and information processing interpretations of phasic cardiac responses.

It has been suggested that cardiac period should be invariably expressed in cardiac time units and that rate should always be expressed in real time units: only these two measures correctly estimate common parameters such as the arithmetic mean (Graham 1978a). However, other arguments may influence the choice between cardiac and real time units: when cardiac activity is time locked to some event, ie, gradually develops during a fixed time period before or after an event independent of the number of intervals between successive R-waves of the electrocardiogram (R-R intervals), it may be more informative to use real time units (Jennings et al 1981).

With the application of an inverse (non-linear) transformation, the raw (ie, observed) heart period data are represented in heart rates, ie, beats per minute (bpm). The non-linear relationship implies that a constant increase in R-R interval at different cycle lengths does not result in a constant decrease in heart rate (Kachaturian et al 1972). At a higher heart rate level a relatively smaller change in period is needed (compared to a lower heart rate level) to establish the same change in beats per minute and vice versa. Because of the inverse relationship, a cardiac rate measure and a cardiac period measure cannot have the same linear relationship with a third variable.

Interpolation of cardiac period into cardiac rate can be carried out in different ways. Commonly, for a particular time interval unit, the number of whole and fractional beats per time unit weighted proportionally is calculated and transformed to beats per minute (Graham 1978a). There has been some dispute about the advantages of either period or rate measures: the use of period measures has been recommended, mainly because periods constitute the raw data set and have somewhat more normal distributions (Kachaturian et al 1972). But according to Graham (1978b), there are neither practical nor statistical advantages for period over rate. When between-group comparisons are made, or when cardiac responses are averaged across trials or states, or when cardiac activity is studied relative to some pre-stimulus level, the effects of the transformation from period to rate ought to be studied carefully. Preferably, both the effects of experimental conditions on heart rate and heart period should be analysed.

Although complex analysis techniques of heart rate like spectral analysis (see Chapter 14) are available, analysis of variance is the most commonly applied technique for analysing phasic heart rate responses. For phasic heart rate studies, change in cardiac rate is often of primary concern, and simple difference scores relative to pre-trial levels are calculated. Effects on heart rate over successive time intervals or beats can be tested with a repeated measures analysis of variance. This analysis method can be applied, for example, with second-by-second changes in heart rate in the interstimulus interval of a reaction time task. The method assumes homogeneous population variances and covariances for each level of a repeated measures factor. This assumption is usually violated. One simple solution for this problem is the Greenhouse and Geisser correction of degrees of freedom (cf Richards, 1980). Not only heart rate patterns, but also selected cardiac response components are analysed, such as peak acceleration and peak deceleration relative to some pre-stimulus level or to a preceding peak measure in a particular time interval (Bohlin and Kjellberg 1979), or cardiac amplitudes at selected time points for which averaged acceleration or deceleration is maximal. The latter, more conservative procedure, emphasises the fixed timing and event dependency of peak accelerations and peak decelerations.

In general, standard criteria for the selection and analysis of phasic heart rate are lacking. Simple recommendations include the development of a general data analytic scheme prior to the experiment which is applied blindly, and the performance of initial data exploration of base levels. When difference scores are used, it is important to check that pre-stimulus heart rate activity does not vary with experimental conditions. Finally, it seems advisable to only interpret task-related differential heart rate changes that are relatively robust and that have been replicated in a different laboratory setting.

SPECIFIC PHASIC HEART RATE RESPONSES

Several reliable phasic cardiac responses and response patterns have been identified in signalled and unsignalled stimulus situations. Three relatively reliable unsignalled cardiac responses that differ in direction, stimulus relatedness and habituation, have been distinguished (Graham 1979). The best known response is orienting response: a small and brief deceleration in response to a novel stimulus or to stimulus change that habituates rapidly. Orienting has been associated with stimulation that carries information and with enhancement of sensory input (Graham 1979). A non-habituating orienting response seems to occur in signalled stimulus situations just after the warning stimulus. Figure 2 illustrates this type of orienting response. The data are from a study performed in our laboratory in which ten loud and aversive 115 dBA tones that lasted two seconds were presented to 15 subjects (Doornen et al 1980). Half of the tones were preceded by a warning signal. At warning a clock hand started and traversed the left part of a clock in 12 seconds. After this period the aversive tone was presented. In this way the aversive tone could be exactly predicted. The other five tones were presented without

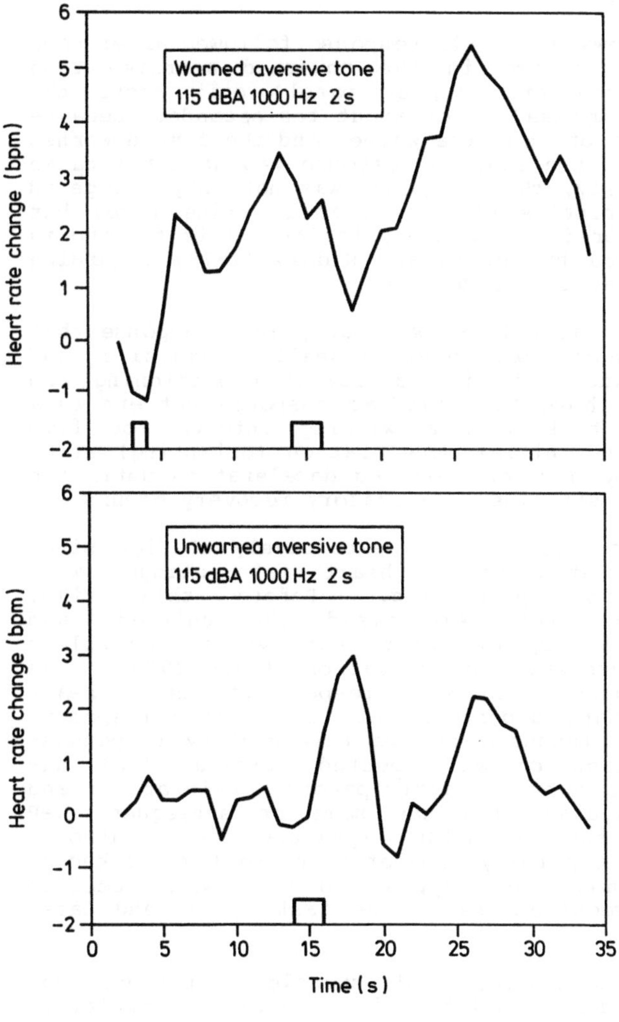

Fig 2. Averaged phasic heart rate change (relative to pre-stimulus level) over ten presentations of an aversive tone of 115 dBA that lasted two seconds. Tones were either preceeded by a warning signal (upper panel) or were presented without warning (lower panel). In the warned condition the inter-stimulus interval was 12 seconds. During this interval a clock hand indicated the amount of elapsed time.

warning. Signalled and unsignalled tone presentations were randomised. The figure shows a small deceleration, interpreted as orienting, after warning onset. Similar orienting responses are found to unsignalled low and moderate intensity tones, to a conditioned stimulus (CS) in classical conditioning studies and to the warning stimulus in signalled reaction time tasks (Bohlin and Kjellberg 1979).

A second response is defense. A defensive response is induced by intense auditory or electric stimulation that is stressful for the subject. The response is related to a reduction of perceptual sensitivity to external stimulation, its amplitude increases with stimulus repetitions, and it is indexed by cardiac acceleration. The third responses, startle, also induces cardiac acceleration but with much shorter latency and habituates with stimulus repetitions. Startle seems to depend on the very rapid transient that occurs at the onset of an auditory stimulus with a rise time of a few milliseconds (Graham 1979).

Figure 2 (lower panel) shows a startle response followed after some seconds by a defensive response in the unwarned aversive tone condition. The warned condition (upper panel) only shows the defensive reflex which is maximal after about ten seconds. Because of the random presentation of the five warned and the five unwarned tones, it is unlikely that the startle response was absent because of habituation. Most likely, this response was not only triggered by the physical characteristics of the stimulus (rise time) but also by its sudden and unpredictable presentation. This finding is in accordance with critique by Turpin and Siddle (1978) regarding the distinction between startle and defense.

A fourth response type is the well known four-phasic response that is observed in repeatedly presented signalled aversive and non-aversive stimulus situations like classical conditioning and reaction time. Figure 3 shows this cardiac response pattern in a signalled reaction time task with a warning interval of five seconds. The figure shows initial deceleration (orienting) after warning onset, followed by acceleration and deceleration until the second stimulus; after the stimulus acceleratory recovery occurs.

This pattern of cardiac change has been extensively studied since the Laceys (1970) associated phasic heart rate changes with attention and information processing. Before this time, interpretations of heart rate emphasised the unified, and energetical aspects of the response: heart rate was viewed solely as an index of general arousal and activation (Malmo 1959). The Laceys criticised this conception with respect to phasic heart rate, and stressed the task dependency of the cardiac response. Their investigations were focussed on two deceleratory responses: one occurs in anticipation of an expected stimulus that has information value for the subject (anticipatory deceleration) and the other is under immediate (in the same or subsequent R-R interval) control of the stimulus (primary deceleration). Anticipatory and subsequent primary deceleration are time locked to the stimulus and their amplitudes appear to vary with specific perceptual and response requirements of the task (Lacey and Lacey 1974).

Unlike the orienting, defensive, and startle responses, the four-phasic pattern in fact consists of a number of different

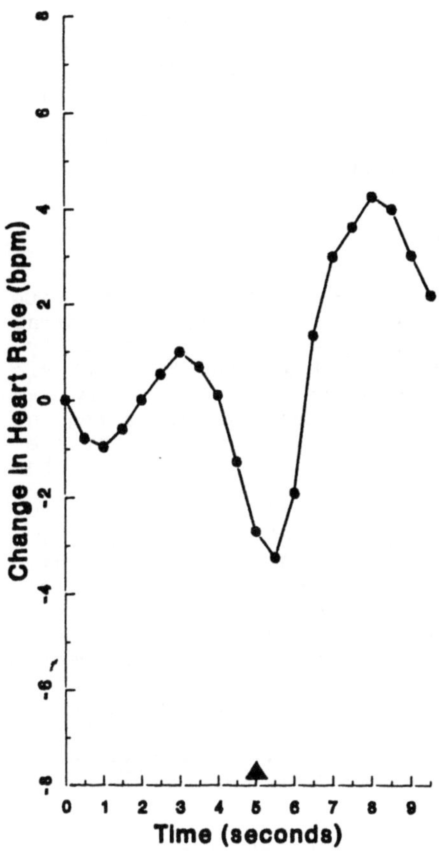

Fig 3. Averaged heart rate change (relative to pre-stimulus level) in a warned choice reaction time task in which a rapid motor response was required. The interval between the warning signal and the imperative signal was five seconds. The figure shows a four phasic response: initial deceleration followed by acceleration and deceleration until the presentation of the imperative stimulus. After this stimulus acceleratory recovery occurred.

task-specific response patterns. The occurrence and amplitudes of the acceleratory and deceleratory components vary with different task situations. For example, there is ample evidence that in classical conditioning tasks, deceleration in the conditioned stimulus-unconditioned stimulus (CS-UCS) interval is relatively enhanced, while acceleration is weakened or even absent (Obrist 1968; Somsen et al 1983). In contrast, shock-avoidance reaction time tasks show both enhanced acceleration and deceleration in the interstimulus interval (Somsen et al 1983; Obrist et al 1979). Thus, as Lang et al (1978) stated, it seems that the psychological meaning of phasic heart rate responses may not be constant across all laboratory paradigms: its significance has to be evaluated in terms of the stimulus context in which it is obtained.

Although the association of phasic heart rate responses and particular information processes is still a matter of debate, it has been firmly established that a task or event which has relevance for the subject induces small but reliable concomitant phasic changes in heart rate. Any subgroup of ten subjects or more that is exposed to a controlled signalled stimulus-response situation shows the four-phasic averaged heart rate response pattern, irrespective of specific laboratory conditions. This is true in spite of the fact that the separate individual response patterns which together constitute the final averaged pattern are often very different (Somsen et al, 1985). The phasic heart rate response appears to result from at least two different influences: a unique individual-specific influence and a task-specific influence that is under control of the ongoing task.

AUTONOMIC CONTROL OF PHASIC HEART RATE

Obrist and his co-workers (1979) reported that heart rate acceleration and deceleration in the CS-UCS interval of a classical conditioning task is controlled by changes in vagal activity. Pharmacological blockade of vagal activity eliminated the biphasic cardiac response under conditions of moderate shock intensity. A small acceleration where heart rate was previously decelerating was found under intense shock conditions. Thus vagal control appeared to dominate the phasic heart rate response, and actually masked sympathetic excitation in the most intense UCS condition.

A similar result was found in a classical conditioning study with dogs. Beta-blockade influenced heart rate only minimally, while the contractile state of the heart was substantially reduced after sympathetic blockade. For reaction time tasks consistent results were found: heart rate deceleration which occurred in anticipation of the imperative stimulus under control conditions disappeared after vagal blockade. Obrist has argued that under non-aversive and passive aversive conditions the phasic cardiac response is under control of parasympathetic activity. Only when subjects are provided some control in the avoidance of aversive stimulation may sympathetic influences on heart rate become manifest (cf Obrist et al 1979).

Animal studies involving direct stimulation of the vagal and sympathetic nerves have consistently demonstrated that vagal influences generally dominate and override sympathetic influences on heart rate (Levy and Martin 1979). Moreover, sympathetic stimulation influences heart rate with a delay of two to three seconds, while the vagus exerts its influence on heart rate almost immediately. In humans, relatively small increases in sympathetic activity may occur during the anticipation period of classical conditioning (Obrist et al 1979) and reaction time tasks (Jennings 1982). However, these sympathetic changes are usually completely masked: a clear phasic deceleration consistently occurs in anticipation of the stimulus in these tasks.

A similar pattern of results has been reported for more cognitive task situations. Ulrych (1969) found similar increases in mean arterial pressure, cardiac output, blood pressure and total peripheral resistance during conversation and mental arithmetic. Tonic heart rate significantly increased in both situations, but acceleration was much greater during mental arithmetic than during conversation. Application of a beta-blocking agent reduced the cardiovascular responses but did not influence heart rate acceleration during conversation. Ulrych suggested that heart rate acceleration during conversation may have been mainly due to release of vagal tone. Unfortunately, the relative contribution of vagal inhibition and sympathetic activation in heart rate was not studied for the mental arithmetic task. However, the pattern of responses suggests that sympathetic excitation increased to the same extent during conversation and mental arithmetic, while the contribution of vagal inhibition was relatively greater for the latter task.

Mental arithmetic tasks have been widely employed in psychophysiology, especially in studies that aimed to investigate the effects of mental stress. In addition to tonic heart rate acceleration and elevations in sympathetic cardiovascular activity, complex mental tasks involving stimulus perception, memorising, symbolic transformation, and overt responding may induce second-by-second changes in heart rate that are closely time locked to the requirements of the task (cf Molen et al, in press). We can speculate that changes in vagal activity may significantly contribute to these temporal changes in heart rate. However, direct evidence about the pattern of sympathetic and parasympathetic influences on phasic heart rate during higher cognitive task performance is lacking.

From the finding that phasic heart rate is predominantly controlled by vagal changes in perceptual-motor and conditioning tasks, it follows that the cardiac response may be secondary to vagally controlled respiratory changes and baroreceptor-mediated changes in blood pressure. Respiration frequency and amplitude have been observed to decrease before the imperative stimulus, but studies involving different methods of controlling respiration (sustained expiration, normal paced, fast paced, and uncontrolled respiration) all showed the typical deceleratory response in anticipation of the imperative stimulus (Porges and Raskin 1969; Obrist et al 1979). In addition, Obrist reported that beta-blockade completely suppressed pressor responses in the CS-UCS interval, but left anticipatory deceleration unaltered. Only acceleration in the interstimulus interval was correlated with inspiration.

Because of vagal mediation, and the finding that the response was not altogether determined by baroreceptor-mediated changes in blood pressure and respiration, Obrist and his co-workers suggested that cardiac deceleration must be controlled by changes in central nervous system activity. They repeatedly observed that anticipatory deceleration is accompanied by inhibition of irrelevant somatic activity as indicated by spontaneous bursts of electromyographic (EMG) activity, chin EMG, respiratory rate, eye movements and blinks. Therefore, it was suggested that at a central level some integration must exist between the cardiac and somatic system, such that they covary during the foreperiod of a reaction time or conditioning task. However, the cardiac response was not considered secondary to these small and irrelevant somatic adjustments: heart rate deceleration was viewed by Obrist as one of the visceral components of the total organismic response in any given situation.

Central mediation (ie, attentional, perceptual and response-related) interpretations of phasic heart rate have been specifically advocated by the Laceys. Bohlin and Kjellberg (1979) have reviewed most of the experimental results and the various attentional and information processing interpretations of the phasic cardiac change.

CARDIAC CYCLE TIME EFFECTS ON PHASIC HEART RATE

Currently, it is widely accepted that phasic heart rate in

behavioural tasks is primarily triggered by parasympathetic influences. Detailed models of how the phasic cardiac response is triggered by vagal excitation and inhibition have been proposed by the Laceys (1978), Jennings and Wood (1977) and Coles et al (1983). It is known that the effect of vagal activity on heart rate depends on the timing of vagal impulses in the cardiac cycle. Animal studies show that as the time from the beginning of the P-wave of the EKG to the vagal stimulus is progressively increased, there is first a progressive lengthening of the same P-P interval, then a rapid decrease and finally a gradual augmentation. This cardiac cycle time effect is greater with more intense vagal stimulation. Moreover, efferent vagal activity normally does not occur in a uniform or randomly dispersed pattern, but there is a tendency toward grouping of impulses in each cardiac cycle (Levy and Martin 1979).

The temporal dependency of vagal activity on heart rate has been attributed to the action of acetylcholine released at the parasympathetic terminals within the sinoatrial node subsequent to vagal stimulation. The Laceys (1978) first demonstrated cardiac cycle time effects in human phasic heart rate responses. They reported that meaningful stimuli that occur early in the cardiac cycle (relative to the R-wave) prolong the same cycle more than late occurring stimuli. Thus, the amplitude of stimulus-induced primary deceleration depends on the timing of stimulus onset within the cardiac cycle. Cardiac cycle time effects were observed for reaction time stimuli, stimuli that had to be detected and counted, and for motor responses that initiated a stimulus. They did not observe cardiac cycle time effects for warning stimuli. Moreover, slow heart rate subjects showed relatively greater cardiac cycle time effects compared to fast heart rate subjects. It was argued that only stimuli and responses that are relevant for the subject show cardiac cycle time effects. Furthermore, the Laceys hypothesised that these stimuli and responses act like direct vagal impulses to the heart: they induce primary bradycardia, the amplitude of which depends on the position of the stimulus or response in the cardiac cycle.

Other studies have elaborated the Laceys findings. Figure 4 shows cardiac cycle time results in a disjunctive reaction time task with foreperiods of 2.5 and 4.5 seconds collected in our laboratory (Molen et al 1983). Cardiac cycle time effects were obtained (as the Laceys did) by post hoc categorisation of trials in which stimulus onset occurred in one of five equally spaced parts (quintiles) of the R-R interval. The upper curves show change in interbeat interval when the subjects actually performed a rapid motor response. The middle curves show the results of catch trials, ie, when the nature of the imperative stimulus indicated that no motor response was allowed. The lower curves show the results for trials in which the warning signal informed the subject in advance that no motor response was required at the imperative stimulus. Thus, in two conditions a motor response was prepared; this response was either executed (upper curves) or inhibited (middle curves) and in one condition no motor preparation was needed (lower curves). The results show that cardiac cycle time effects occurred when stimuli provided information about the task (with or without a motor response), while signalled control stimuli without task relevance did not induce cycle time effects.

Another cycle time effect was investigated by Jennings and Wood (1977). They studied cardiac cycle time effects by presenting

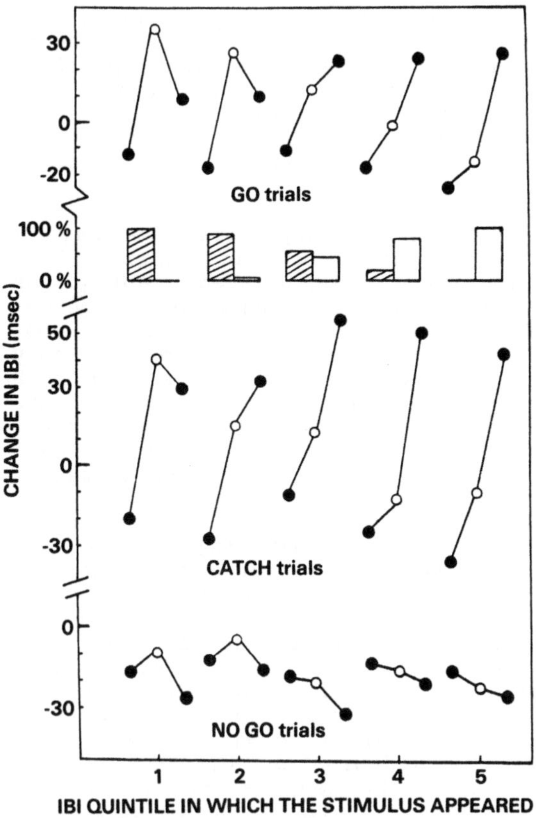

Fig 4. Change in interbeat interval (IBI) proximal to the imperative stimulus in a disjunctive reaction time task as a function of the preceding R-wave of the EKG to the tone. The preceding (-1) R-R interval, the R-R interval of stimulus presentation (0), and the succeeding R-R interval (+1) are depicted in each curve. The differences of IBIs and prewarning baseline are presented for each quintile of the R-R interval in which the stimuli occurred. The shaded bars indicate the proportion of responses that occurred in R-R interval 0 and the open bars the proportion of responses that occurred in R-R interval +1. Interval -1 and +1 are arbitrarily displaced along the abscissa over an equal distance in either direction from the stimulus R-R interval. From Molen et al, 1983.

stimuli at fixed time delays (0 and 350 msec) from the R-wave. They designed a hypothetical model of the time course of vagal activity to the heart. According to this model the greatest accumulation of vagal impulses to the heart occurs between 200 and 400 msec after the R-wave. Jennings further hypothesised that response completion in a reaction time task stops vagal activity, ie, induces vagal inhibition and consequently triggers a shift from deceleration to acceleration. The results showed that responses occurring before 300 msec after the R-wave induced the same beat to shift to acceleration, while responses that occurred more than 300 msec after the R-wave did not affect the same beat interval (deceleration continued) but induced the subsequent beat interval to shift to acceleration.

RELEVANCE OF PHASIC HEART RATE MEASUREMENT

The evidence discussed so far does not indicate that phasic heart rate changes have much relevance for the field of stress and coronary heart disease or other research of individual differences. As Obrist et al (1978) stated, a traditional psychosomatic hypothesis implicitly assumes that cardiac changes in response to external events must largely involve sympathetic influences. The clear parasympathetic control of phasic heart rate argues against this hypothesis. This observation illustrates that attention in many stress studies has been predominantly devoted to sympathetic responses to stress, and that vagal effects and possible interaction of vagal and sympathetic activity during stress have not been considered to be of much relevance. In the next section some results will be discussed which indicate the potential relevance of vagally induced changes of heart rate in stress studies.

It is known from animal studies that the vagal response may be enhanced when sympathetic tone is increased (Levy and Martin 1979). Systematic investigations of this accentuated antagonism phenomenon showed that stimulation of the sympathetic nerves increased heart rate considerably when the vagi were inactive but had only negligible effects when the vagi were simultaneously stimulated at a constant rate. The reverse effect was even more potent. Increased stimulation of the vagi decreased heart rate considerably in the absence of sympathetic tone. The same increase in vagal stimulation may induce almost twice the reduction in heart rate when the sympathetic nerves are stimulated at a constant rate.

Accentuated antagonism in humans was reported by Carruthers and Taggart (1973) who studied the effect of watching violent and suspense film scenes on tonic heart rate and plasma adrenaline. In one group of subjects sympathetic innervation of the heart was pharmacologically (beta) blocked. Adrenaline secretion doubled in the control group and increased less markedly in the blocked group, while heart rate slowing remained unchanged. Marked deceleration occurred during all film scenes. Deceleration was greater during violent and suspense scenes compared to earlier and later control scenes. Change in vagal influence on heart rate was estimated from the sinus arrhythmia gap, ie, the widening of the gap between maximum and minimum heart rate during the respiratory cycle. The violent and suspense scenes induced an increase in maximum heart rate and a decrease in minimum heart rate. It appeared therefore that an increase in vagal influence on heart rate occurred. However, the net result on heart rate of increased sympathetic and vagal activity during the violent and suspense scenes of the films was increased heart rate deceleration. These results clearly indicate that vagal task- and event-related changes in heart rate may interact with sympathetic changes in stressful task situations.

Figure 5 illustrates an example of the interaction between individually experienced stress and task-related phasic activity in heart rate. The data in this figure were collected in our laboratory as part of research conducted by van Doornen. The figure shows the heart rate responses of two participants to a public thesis defense meeting: the defender (curve with highest heart rate level) and the supervising professor. Heart rate was monitored from about 15 minutes before until about 20 minutes after the meeting. At 80 minutes the questioning was over, and at about 90 minutes the supervising professor made a formal speech before handing down the PhD.

It appears that three different types of stress-related heart rate responses are present in these subjects. Firstly, sustained heart rate acceleration that gradually decreased during the session and that probably results from increased humoral sympathetic tone (defender). Similar patterns of prolonged tachycardia together with increased excretion of catecholamines have been reported in subjects speaking before an audience (Taggart et al 1973).

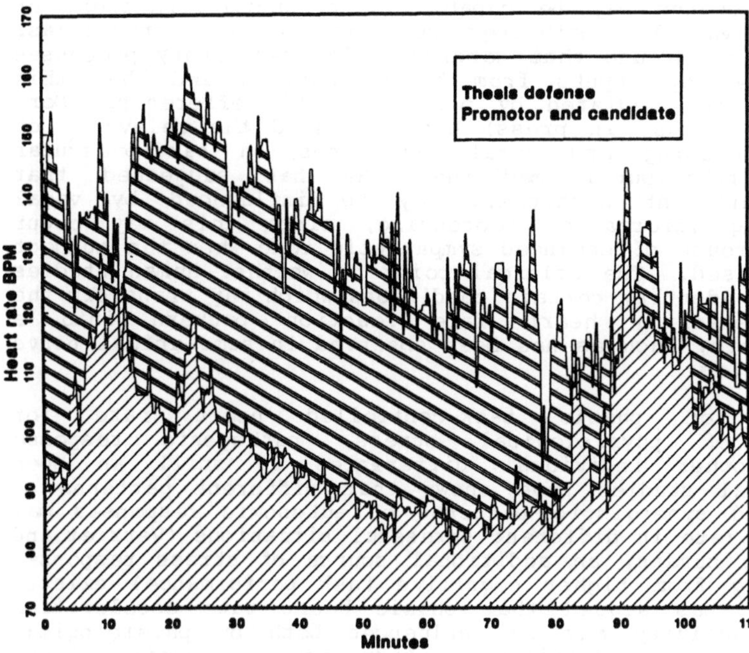

Fig 5. Phasic heart rate of two subjects before (15 minutes), during (75 minutes) and after public doctoral thesis defense meeting. Heart rate of the defender of the thesis (upper curve) and the supervising professor (lower curve) are depicted. At about 25, 35, 45, 65 minutes after the beginning of the recordings one of five experts started his interrogation which was subsequently answered by the defender. After about 90 minutes the supervising professor made a formal speech and the candidate received the PhD.

Secondly, bursts of cardiac acceleration that lasted several minutes occurred. These bursts occurred before and just after the beginning of the meeting and during the speech in the professor's heart rate. The same type of burst-like elevations in heart rate have been reported by Wenger et al (1960) after infusions of adrenaline and noradrenaline. Finally, phasic heart rate changes that closely mirror the ongoing events were shown by the candidate, who was actively listening to the questions and subsequently answered them. The event-related phasic heart rate changes of the professor (who was only passively listening) were much smaller. The phasic changes of the candidate sometimes mounted up to 30 bpm within a single minute. Based on the previous discussion, we can speculate that these increased event-related heart rate changes may have been largely controlled by short-term fluctuations in vagal activity.

This interpretation indicates the potential relevance of exaggerated vagal effects on phasic heart rate in stressful situations. Cohen and Obrist (1975) have discussed how cardiovascular responses (normal or pathological) may occur in anticipation of behavioural demands, and how repeated occurrences of particular demands may be capable of producing transient or sustained cardiovascular pathology such as cardiac arrhythmias. Steptoe (1985) has argued that cardiovascular regulatory processes may be modulated by inputs from higher centres and that this modulation is often mediated by parasympathetic efferents. More specifically, Jennings (in press) has discussed the relevance of information processing and vagal influences on cardiovascular function for behavioural medicine. He has suggested that clinically significant arrhythmia may be influenced by vagal responses during information processing, since this may occur against a background of enhanced sympathetic tone when the subject is somewhat aroused. The critical point is whether human studies will be able to identify certain conditions which may transform the normal non-threatening heart rate reaction pattern into a pathogenic pattern, such as has been observed in some animal nerve stimulation studies.

The relevance of phasic task-controlled heart rate for investigations of physiological stress, pathogenic response patterns and other individual differences is still an open question. Animal studies have shown that different patterns of antagonistic and synergistic patterns of autonomic responses may influence heart rate in different stressful situations (Verrier and Lown 1978). In humans, an additional source of stress responses (those resulting from higher cognitive task performance) exists. Jennings (1983) has argued that phasic heart rate acceleration curing mental activity may be influenced both by physiological stress responses and cognitive information processing. Our knowledge of such complex determinants of phasic heart rate change is, however, very limited. The optimal strategy for experimental studies which aim to investigate effects of individual differences and stress in phasic heart rate in tasks involving higher mental activity, seems to be to study both the task-specific and individual-specific effects on the phasic cardiac response, and to relate these to the underlying pattern of vagal and sympathetic change.

ACKNOWLEDGEMENTS

I wish to thank Lorenz van Doornen and Ko Orlebeke for their critical reading of this manuscript.

Preparation of this manuscript was supported by a grant from the Society of Pure Scientific Research in the Netherlands.

REFERENCES

Bohlin G, Kjellberg A (1979) Orienting activity in two stimulus paradigms as reflected in heart rate. In: Kimmel HD, van Olst EH, Orlebeke JF (eds) The Orienting Reflex in Humans. Lawrence Erlbaum Associates, Hillsdale New Jersey, p169

Carruthers MM, Taggart P (1973) Vagotonicity of violence: biochemical and cardiac responses to violent films and television programmes. Brit Med J 3: 384-389

Cohen H, Obrist P (1975) Interactions between behavior and cardiovascular system. Circulation Res 37: 693-706

Coles MGH, Pellegrini AM, Wilson GV (1983) The cardiac cycle time effect: influence of respiration phase and information processing requirements. Psychophysiology 19: 648-657

Doornen LJP van, Orlebeke JF, Somsen RJM (1980) Coronary risk and coping with aversive stimuli. Psychophysiology 17: 598-603

Graham FK (1978a) Constraints on measuring heart rate and period sequentially through real and cardiac time. Psychophysiology 15: 492-495

Graham FK (1978b) Normality of distributions and homogeneity of variance of heart rate and heart period samples. Psychophysiology 15: 487-491

Graham FK (1979) Distinguishing among orienting, defense and startle reflexes. In: Kimmel HD, van Olst EH, Orlebeke JF (eds) The Orienting Reflex in Humans. Lawrence Erlbaum Associates, Hillsdale New Jersey, p137

Jennings JR (1982) Beat-by-beat vascular responses during anticipatory heart rate deceleration. Physiol Psychol 10: 422-430

Jennings JR (1983) Attention and coronary heart disease. In: Krantz DS, Baum A, Singer JE (eds) Handbook of Psychology and Health, Vol III, Cardiovascular Disorders and Behavior. Lawrence Erlbaum Associates, Hillsdale New Jersey, p85

Jennings JR (in press) Could vagal responses during information processing be related to clinically significant arrhythmia? In: Twentyman C, Epstein L, Blanchard E, Brady JV (eds) Progress in Behavioral Medicine. Springer Verlag, New York

Jennings JR, Wood CC (1977) Cardiac cycle time effects on performance, phasic cardiac responses, and their intercorrelation in choice reaction time. Psychophysiology 14: 297-307

Jennings JR, Berg WK, Hutcheson JS, Obrist P, Porges S, Turpin G (1981) Publication guidelines for heart rate studies in man. Psychophysiology 18: 226-231

Kachaturian ZS, Kerr J, Kruger R, Schachter J (1972) A methodological note: comparison between period and rate data in studies of cardiac function. Psychophysiology 9: 539-545

Lacey JI, Lacey BC (1970) Some autonomic-central nervous system interrelationships. In: Black P (ed) Physiological Correlates of Emotion. Academic Press, New York, p205

Lacey JI, Lacey BC (1974) Studies on heart rate and other bodily processes in sensorimotor behavior. In: Obrist PA, Black AH, Brener J, DiCara LV (eds) Cardiovascular Psychophysiology. Aldine, Chicago, p538

Lacey BC, Lacey JI (1978) Two-way communication between the heart and the brain: significance of time within the cardiac cycle. Amer Psychol 33: 99-113

Lang PJ, Ohman A, Simons RF (1978) The psychophysiology of anticipation. In: Requin J (ed) Attention and Performance VII. Lawrence Erlbaum Associates, Hillsdale New Jersey

Levy MN, Martin PJ (1979) Neural control of the heart. In: Berne RM, Sperelakis N, Geiger SR (eds) Handbook of Physiology, Section 2, The Cardiovascular System. Geiger, Bethesda Maryland: American Physiological Society, p581

Malmo RB (1959) Activation: a neuropsychological dimension. Psychol Rev 66: 367-386

Molen MW van der, Somsen RJM, Orlebeke JF (1983) Phasic heart rate responses and cardiac cycle time in auditory choice reaction time. Biol Psychol 6: 255-272

Molen MW van der, Somsen RJM, Orlebeke JF (in press) The rhythm of the heart beat in information processing. In: Ackels PK, Jennings JR, Coles MGH (eds) Advances in Psychophysiology. JAI Press, Greenwich

Obrist PA (1968) Heart rate and somatic motor coupling during classical aversive conditioning in humans. J Exper Psychol 77: 180-193
Obrist PA, Light KC, Langer AW, Grignolo A, McCubbin JA (1978) Behavioral-cardiac interactions: the psychosomatic hypothesis. J Psychosom Res 22: 301-325
Obrist PA, Langer AW, Light K, Grignolo A, McCubbin JA (1979) Myocardiac performance and stress: implications for basic and clinical research. In: Kimmel HD, van Olst EH, Orlebeke JF (eds) The Orienting Reflex in Humans. Lawrence Erlbaum Associates, Hillsdale New Jersey, p219-240
Porges SW, Raskin DC (1969) Respiratory and heart rate components and attention. J Exper Psychol 81: 497-503
Richards JE (1980) The statistical analysis of heart rate: a review emphasizing infancy data. Psychophysiology 17: 153-166
Somsen RJM, Molen MW van der, Orlebeke JF (1983) Phasic heart rate changes in reaction time, shock avoidance, and unavoidable shock tasks: are hypothetical generalizations about different S1-S2 tasks justified? Psychophysiology 20: 88-94
Somsen RJM, Molen MW van der, Boomsma DI, Orlebeke JF (1985) Phasic cardiac responses in reaction time and mental arithmetic tasks: the dominant influence of mental task performance on heart rate in adolescents. In: Orlebeke JF, Mulder G, van Doornen LJP (eds) Psychophysiology of Cardiovascular Control: models, methods and data. Plenum Press, New York
Steptoe A (1985) Behavioral influences of the mechanisms of hemodynamic regulation. In: Orlebeke JF, Mulder G, van Doornen LJP (eds) Psychophysiology of Cardiovascular Control: models, methods and data. Plenum Press, New York, p169
Taggart P, Carruthers M, Somerville W (1973) Electrocardiogram, plasma catecholamines and lipids, and their modificaton by oxprenolol when speaking before an audience. Lancet ii: 341-346
Turpin G, Siddle DA (1978) Cardiac and forearm plethysmographic responses to high intensity auditory stimulation. Biol Psychol 6: 267-281
Ulrych M (1969) Changes of general haemodynamics during stressful mental arithmetic and non-stressing quiet conversation and modification of the latter by beta-adrenergic blockade. Clin Sci 36: 453-461
Verrier RL, Lown B (1978) Sympathetic-parasympathetic interactions and ventricular electrical stability. In: Schwartz PJ, Brown AM, Malliani A, Zanchetti A (eds) Neural Mechanisms in Cardiac Arrhythmias. Raven Press, New York, p75
Wenger MA, Clemens TL, Darsie ML, Engel BT, Estes FM, Sonnenschein RR (1960) Autonomic resonse patterns during intravenous infusion of epinephrine and norepinephrine. Psychosom Med 12: 294-307

CHAPTER 14

Measurement and Analysis of Fluctuations in Heart Rate and Blood Pressure

Otto Rompelman

Information Theory Group, Technische Hogeschool Delft, Afdeling der Elektrotechniek, PO Box 5031, NL-2600 GA DELFT, The Netherlands

SUMMARY

The analysis of fluctuations in heart rate and blood pressure may provide useful information on the activity of the autonomic nervous system. These fluctuations are defined in terms of time series or signals. Different ways of defining heart rate variability and blood pressure fluctuations are reviewed. It is shown how different definitions may lead to different interpretations of the same data. The concept of spectral analysis is introduced and two possible methods for spectral analysis of heart rate variability are discussed.

INTRODUCTION

Two important cardiovascular parameters that are often used in the field of experimental psychology are heart rate and blood pressure. As is the case with all physiological quantities, these parameters are subject to variations resulting from both internal and external factors. We may differentiate between slow and fast variations, the latter ones being mainly generated by neural activity. Roughly speaking, the time course of slow variations is in the order of minutes, whereas the fast variations take place in seconds or less. Of course, there is an overlap between the two. The measurement of these variations, however, requires two different approaches. In this contribution the differences between the approaches are briefly discussed, together with some practical problems encountered when analysing the results.

MEASUREMENT OF SLOW VARIATIONS

The measurement of slow variations does not necessarily require sophisticated instruments. Blood pressure can be measured by means of the traditional cuff method. If we take measurements over, for example, ten minutes, we obtain a series of values which can be regarded as a time series. The same holds for heart rate. However, since a rate is a quantity based on an averaging procedure it is most important to take the averaging time into account. For instance, we may count the number of heart beats that occurred in a one minute interval or in a ten-second interval. What is the difference between the two methods? First, it is obvious that, if we take the average heart rate over each ten seconds we can follow rapid changes more accurately than when heart rate is based on one minute averages.

But there is another difference. Suppose that there is a periodical variation in heart rate with a duration equal to the measurement (counting) interval: this variation is completely averaged out. This may become a serious problem, since a ten-second averaging interval may severely damp the ten-second rhythm to be discussed later. The same type of difficulty occurs in the measurement of blood pressure. In particular, if we define blood pressure as the mean arterial pressure the word 'mean' implies an averaging procedure and again the question arises: what averaging time should be chosen?

In general, it can be said that the averaging time should be in agreement with the type of phenomenon one is interested in.

MEASUREMENT OF BLOOD PRESSURE VARIABILITY

Variations in heart rate and blood pressure are measured as variance in a mean value. The problem that has to be resolved is the time over which the mean is to be calculated. If we are interested in fast fluctuations we have to shorten this period as much as possible. In practice this implies that we have to use methods that give more or less beat-by-beat information. Obviously a beat-by-beat measurement of heart rate variability or blood pressure variability is the speed limit.

The beat-by-beat analysis may require sophisticated instruments. In the case of blood pressure variability, we need a continuous recording of the intra-arterial pressure obtained either invasively by a catheter system or non-invasively with the help of the Penáz method (eg, van Vollenhoven, in press). The mean arterial pressure can be obtained from the recorded signal by averaging so that the beat-by-beat fluctuations in the mean pressure remain. This means that the signal has to be passed through a low pass filter with a cut-off frequency of about 0.5Hz. The filter should have a high rejection for higher frequencies in order to suppress the blood pressure waveform effectively.

MEASUREMENT OF HEART RATE VARIABILITY

The measurement of heart rate variability also requires an averaging procedure with a time constant such that beat-by-beat variations can be measured. The source for this measurement is the ECG which nowadays can be easily obtained. Two important steps must be carried out before arriving at a heart rate variability signal, namely:

- detection of the QRS-complex;
- conversion of the QRS event series into a heart rate variability signal.

The detection of QRS complexes will not be discussed here. A series of pulses are then derived from the ECG. The four most common methods of converting the event series into a heart rate variability signal will be briefly described (Rompelman et al 1977).

1. The Interval Tachogram (IT), Fig 1. The intervals between the events are given as a function of the interval number.

$Y(i) = [t_i - t_{i-1}]$ t_i = event occurrence times

This plot is very easy to obtain since an event process is usually stored in the form of its intervals. Caution should however be used when this function is subjected to spectral analysis. Since the tachogram is a function of the interval number rather than time an interval spectrum is obtained, the independent variable (abscissa) being expressed in cycles/interval rather than cycles per second (or Hertz).

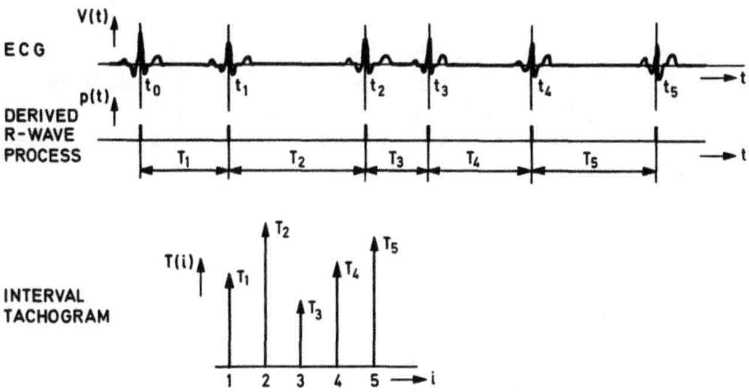

Fig 1. The ECG and its derived RR-interval tachogram

2. The Interval Function (IF), Fig 2. The intervals between the events are given as a function of time:

$$Y(t) = \sum_{i=-\infty}^{\infty} (t_i - t_{i-1}) . \delta(t - t_i).$$

t_i = QRS occurence time

The IF can be considered as an irregularly sampled waveform. In order to make analysis possible one usually applies some kind of interpolation in order to render a signal suitable for continuous sampling.

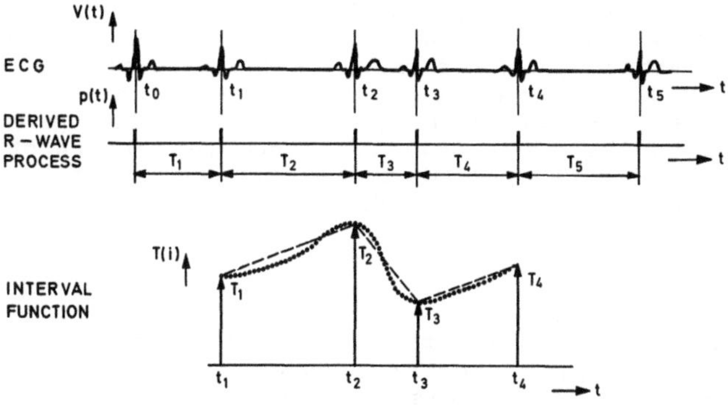

Fig 2. The ECG and its derived interval function; two methods of interpolation are shown

3. The Instantaneous Heart Rate (IHR). This is the reciprocal of the IF, if the instantaneous heart rate is defined as the reciprocal of the cardiac event interval. If this signal is passed through a so-called zero order hold circuit, one obtains the signal which is produced by the cardio-tachometer. Some practical consequences of both the IF and the IHR approach will be discussed later.

4. The Low Pass Filtered Event Series (LPFES). This method is based on a model of the neural influence of the sinoatrial (SA) node, discussed below.

The question arises of which method is to be preferred. Before answering this question it is important to note that whatever method is used, it will be based (implicitly or explicitly) on a model. This model accounts for the total neural influence on the SA-node resulting in fluctuations in heart rate (Fig 3).

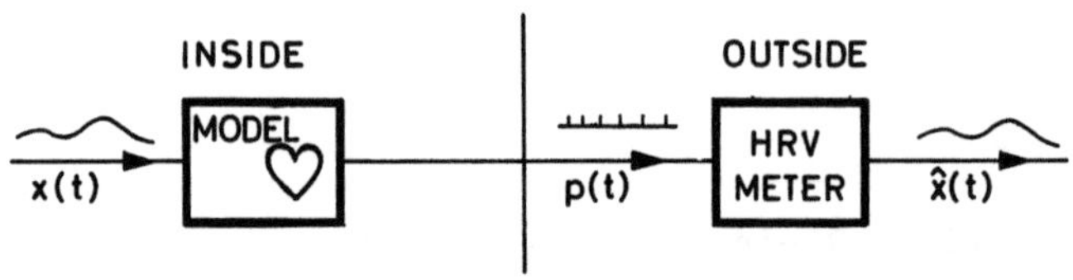

Fig 3. The choice of a particular heart rate variability meter, which converts the cardiac event series p(t) into a heart rate variability signal $\hat{x}(t)$ implies a model, which converts the hypothetical signal x(t) into the cardiac event series x(t) such that $\hat{x}(t)=x(t)$

Hyndman and Mohn (1973) were the first who based the analysis of heart rate variability on a model of the natural pacemaker, including the neurally induced variations in firing rate. This model, the Integral Pulse Frequency Modulator (IPFM) is shown in Fig 4. The input signal m(t), which is supposed to represent the net vagal and sympathetic activity, is integrated, yielding y(t): when y(t) reaches a fixed reference level, R, a pulse is emitted and the integrator is reset to zero whereupon the cycle starts again. The sequence of emitted pulses represents the cardiac event series x(t). Earlier work of Bayly (1968) had shown that the modulating signal m(t) can be regained by means of a low pass filter operation applied to the event series. Since the origin of fluctuations in the rate of occurrence of the events was explained in terms of a <u>modulation</u>, the low pass filtering operation may also be called <u>demodulation</u>.

Hyndman's LPFES is therefore based on this model. The next question is: how do we perform this filtering (or demodulation) in practice? This procedure is in fact rather simple, and can be carried out by means of a simple computer program. An important feature of the demodulator is the impulse response of the filter applied. Filter theory will not be discussed here, but the reader should be reminded that filters with a symmetrical impulse response do not affect the phase relations within the signal. However, a

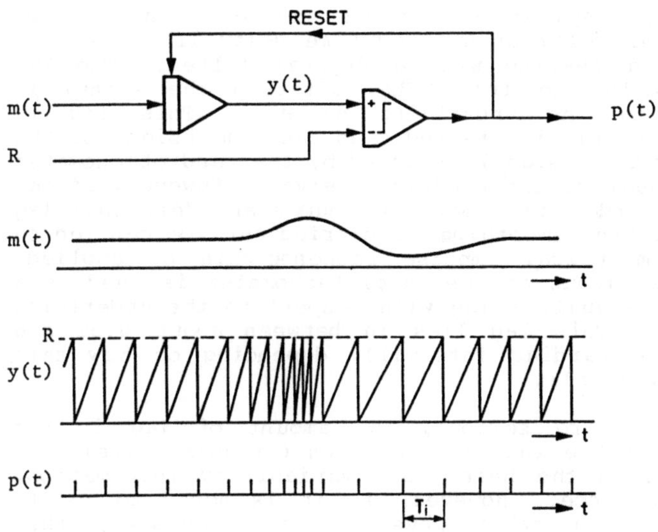

Fig 4. The Integral Pulse Frequency Modulator (IPFM) as a model for sinus node activity

symmetrical impulse response can only be realised off-line. Off-line processing is usually carried out with a digital computer. In Fig 5 an example is shown of the LPFES in relation to respiration. The heart rate variability signal was obtained on-line with the help of a digital filter with a symmetrical impulse response but with a fixed delay (three seconds). More details of the features of this system are described in Coenen et al (1977).

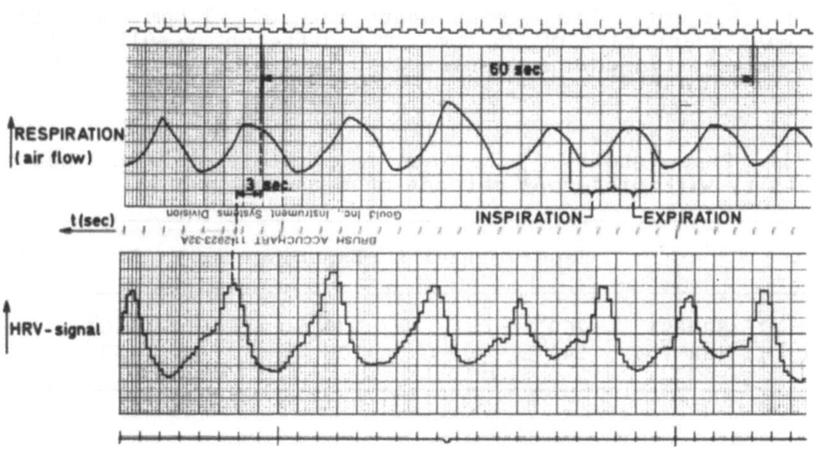

Fig 5. Top: respiratory flow signal. Bottom: simultaneously recorded heart rate variability signal, obtained on-line with a digital filter; note the three second delay due to the time shift inherent in the filter procedure

Provided a filter with a symmetrical impulse response is used, we do not introduce a time delay or a fixed time delay if heart rate variability is measured on-line with a digital filter. The IHR produces a signal similar to the LPFES, although the signal is lagged with respect to the underlying process. This lag is dependent on the way the IHR is treated (and the same holds for the IF). If the rather jumping signal produced by the cardiotachometer is used, the lag is equal to one cardiac interval. However, if the signal is smoothed in order to remove the jumps an additional lag is introduced. If off-line smoothing is carried out, we can reduce the lag because a symmetrical impulse response can be applied. Recapitulating: if the output of the cardiotachometer is used as a heart rate variability signal, a lag with respect to the underlying process is introduced. This lag lies in between about $0.5\overline{T}$ and $1.5\overline{T}$ ($\overline{T}$ is the average cardiac interval), depending on how this signal is processed, if at all.

Moreover, and this is more serious, the amount of lag is not constant but varies with the interval duration (or heart rate). If we are only interested in the heart rate variability information, this is of minor importance. However, if it is hoped to relate heart rate variability to other physiological processes, this becomes very important. As an example, suppose that we want to relate heart rate variability to respiration and that the mean heart rate during the experiment is 75 beats per minute (bpm). A cardiotachometer is used for the measurement of heart rate variability. The lag is then 800ms. If the instrument is triggered from the QRS-complexes we have to add an additional delay, since the actual firing of the SA node takes place about 200 ms before the QRS-derived event.

Let us expand this example more and introduce three experimenters: 'A' uses a QRS-triggered cardiotachometer; 'B' uses a QRS trigger delay compensated low pass filter; and 'C' uses the same instrument and compensates for the SA-QRS delay. All three want to measure the phase shift between respiratory flow and heart rate variability of two subjects breathing at a constant rate of 15 breaths/min (ie, 0.25Hz). Subject I has a mean heart rate of 60 bpm and Subject II has a mean heart rate of 75 bpm. The different results obtained by the three investigators are summarised in Table 1.

Table 1. Phase shift ψ_{rh} between respiratory flow and heart rate variability with different methods of measurement

	Experimenter A	Experimenter B	Experimenter C
Subject I ($\overline{HR}$=60)	260°	170°	150°
Subject II ($\overline{HR}$=75)	240°	170°	150°

It is obvious that a scientific discussion between A, B and C on the physiological processes underlying respiratory arrhythmia will not be very enlightening if the measuring conditions are not taken into account!

The LPFES is the most suitable signal for heart rate variability analysis. A discussion of the different possible realisations of the low pass filter is beyond the scope of this chapter. Needless to say that filters with different impulse responses will yield different results. For some examples of the consequences of different filters, see Rompelman (in press).

SOME REMARKS ON SPECTRAL ANALYSIS

Heart rate variability and blood pressure variability can be described in terms of mean values or variance: this gives information about the level and power of the fluctuations. However, if we want to study the 'internal structure' of the signal, we have to carry out correlational or spectral analysis. The spectrum provides information about the more or less periodical components present in the data (by 'more or less' we imply that spectral analysis need not be confined to pure sine waves, and that small variations in amplitude and/or phase are permitted). The spectrum of a signal x(t) is defined by its Fourier-transform:

$$X(f) = \int_{-\infty}^{\infty} x(t) \, . \, e^{-j2\pi ft} dt = R(f) + jI(f)$$

The power spectrum follows from:

$$S(f) = R^2(f) + I^2(f)$$

A practical problem is the integration from minus to plus infinity. Since we always deal with finite data or with non-stationary data, we can only obtain an estimate of the spectrum. A feature of the spectrum is the spectral resolution. There is an important relation between the spectral resolution B and the segment length T of the data, such that B x T is a constant. Hence, decreasing the number of data points degrades the spectral resolution, or, the other way around: if we require high spectral resolution, a large number of data points is needed.

Figure 6 shows an example of the heart rate variability spectrum of a resting healthy subject. Two important phenomena are apparent. These are the respiratory component, due to cardiorespiratory interaction (RSA), and a ten-second rhythm component (oscillations in the baroreceptor reflex loop). With spectral analysis it is possible to study both the respiratory arrhythmia and the relations between mental load and the ten-second rhythm (see Chapter 3).

There are two ways of obtaining a heart rate variability spectrum:

1. by the Fourier transform of the heart rate variability signal (indirect method) as mentioned above;

2. by direct spectral analysis of the event series (direct method).

This latter method was reported by Rompelman et al (1982) and will be briefly described here.

Assume the cardiac event series x(t) is given as a series of δ-pulses:

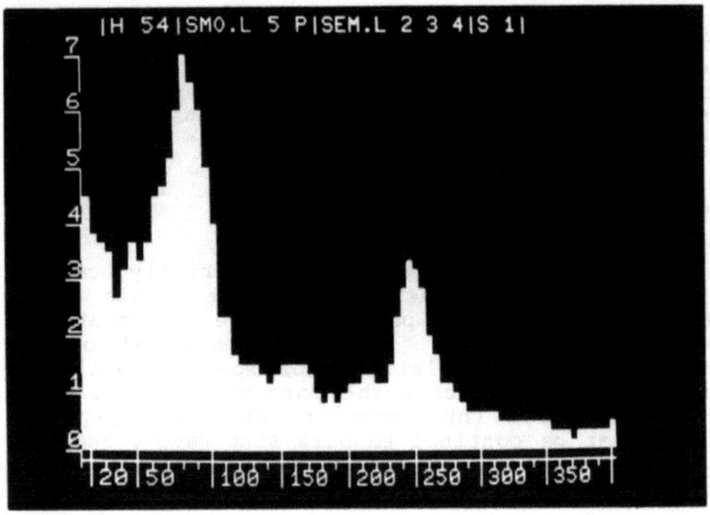

Fig 6. Example of a heart rate variability spectrum (unit of the horizontal axis is mHz); note the ten second rhythm (0.1Hz) and the cardiorespiratory component ($\sim$ 0.25Hz)

$$x(t) = \sum_{i=-\infty}^{\infty} \delta(t-t_i) \qquad t_i = \text{event occurence time}$$

Taking the Fourier transform and elaborating:

$$X(f) = \int_{-\infty}^{\infty} \left[\sum_{i=-\infty}^{\infty} \delta(t-t_i) \right] . \, dt$$

$$= \sum_{i=-\infty}^{\infty} e^{-j2\pi f t_i}$$

Hence, the power spectrum is:

$$S(f) = \left[\sum_{i=-\infty}^{\infty} \cos 2\pi f t_i \right]^2 + \left[\sum_{i=-\infty}^{\infty} \sin 2\pi f t_i \right]^2 .$$

It is evident that the computational effort is very low. A comparison of the indirect and direct methods for heart rate variability spectral analysis is given in Table 2.

CONCLUSION

Some aspects of the analysis of fluctuations in heart rate and blood pressure are reviewed. It has not been the goal of this contribution to repeat all that has been written on this topic, but rather to provide a short introduction to the field. The interested reader is referred to the literature, a selection of which is given below.

Table 2. Comparison of the direct and indirect method of creating a heart rate variability spectrum

	Indirect method	Direct method
On-line spectral analysis	Hardly possible	Very satisfactory
Off-line spectral analysis	Yes	Yes
Calculational complexity	High	Very low
Method based on a model	Yes: depends on creation of heart rate variability signal	Yes: for the same reason
Heart rate variability signal available for further analysis?	Yes	No
Spectral shape dependent on filter used for creation of heart rate variability signal?	Yes	Does not apply

REFERENCES

Bayly EJ (1968) Spectral analysis of pulse frequency modulation in the nervous system. IEEE Transact Biomed Eng BE-15: 257-265

Coenen AJ, Rompelman O, Kitney RI (1977) Measurement of heart rate variability, part 2. Med & Biol Eng & Comp 15: 423-430

Hyndman BW, Mohn RK (1973) A pulse modulator model for pacemaker activity. In: Digest of Xth Int Conf on Med & Biol Eng. Dresden, p233

Rompelman O, Coenen AJ, Kitney RI (1977) Measurement of heart rate variability, part 1. Med & Biol Eng & Comp 15: 233-239

Rompelman O, Snijders JB, van Spronsen CJ (1982) The measurement of heart rate variability spectra with the help of a personal computer. IEEE Trans Biomed Eng BME-29: 503-510

Rompelman O (in press) Investigating heart rate variability: problems and pitfalls. In: Grossman P, Janssen K, Vaitl D (eds), Cardio-respiratory and -Somatic Psychophysiology. Plenum Press, New York

van Vollenhoven E (in press) Measurement of blood pressure. In: Kitney RI, Rompleman O (eds), The Beat-by-beat Analysis of Cardiac Function. Clarendon Press, Oxford

<u>Other References in the Field of Analysis of Heart Rate Variability and Blood Pressure Variability</u>

Automedia (1983) Special issue on analysis of heart rate variability and blood pressure variations. Vol 4

Kitney RI, Rompelman O (1980) The Study of Heart Rate Variability. Clarendon Press, Oxford

Kitney RI, Rompelman O (in press) The Beat-by-beat Analysis of Cardiac Function. Clarendon Press, Oxford

Orlebeke JF, Mulder G, van Doornen LJP (1985). Psychophysiology of Cardiovascular Control. Plenum Press, New York

Other References in the Field of Signal Analysis

Bendat JS, Piersol AG (1980) Engineering Applications of Correlation and Spectral Analysis. John Wiley, New York
Brigham EO (1974) The Fast Fourier Transform. Prentice-Hall, Englewood Cliffs NJ

Subject Index

MIX
Papier aus verantwortungsvollen Quellen
Paper from responsible sources
FSC® C105338

If you have any concerns about our products,
you can contact us on
ProductSafety@springernature.com

In case Publisher is established outside the EU,
the EU authorized representative is:
Springer Nature Customer Service Center GmbH
Europaplatz 3, 69115 Heidelberg, Germany

Printed by Libri Plureos GmbH
in Hamburg, Germany